T0323288

Being Rational and Being Right

Being Rational and Being Right

JUAN COMESAÑA

Great Clarendon Street, Oxford, OX2 6DP,
United Kingdom

Oxford University Press is a department of the University of Oxford.
It furthers the University's objective of excellence in research, scholarship,
and education by publishing worldwide. Oxford is a registered trade mark of
Oxford University Press in the UK and in certain other countries

First Edition published in 2020

Impression: 1

Published in the United States of America by Oxford University Press
198 Madison Avenue, New York, NY 10016, United States of America

British Library Cataloguing in Publication Data
Data available

Library of Congress Control Number: 2019947642

ISBN 978–0–19–884771–7

DOI: 10.1093/oso/9780198847717.001.0001

Printed and bound in Great Britain by
Clays Ltd, Elcograf S.p.A.

To Manuel Comesaña, my first and most influential philosophical mentor

Table of Contents

Acknowledgments xi

1. Introduction 1

2. Probability and Decision Theory 6
   - 2.1 Introduction 6
   - 2.2 Probabilities: An Axiomatic Approach 7
     - 2.2.1 Sets 7
     - 2.2.2 Propositions—and Bayes's Theorem 9
   - 2.3 Probabilities: A Measure-Theoretic Approach 11
   - 2.4 Probabilities: Stochastic Truth-Tables 12
   - 2.5 Problems with Conditional Probability 13
   - 2.6 Physical Probabilities 14
   - 2.7 Normative Probabilities: Probabilism and Conditionalization as Necessary Conditions on Rational Credence 16
     - 2.7.1 Decision under Uncertainty? 17
     - 2.7.2 Probabilism and the Problem of Logical Omniscience 19
     - 2.7.3 Conditionalization 31
   - 2.8 Traditional Decision Theory 33
     - 2.8.1 Actions, States, and Outcomes 33
     - 2.8.2 Utility 34
     - 2.8.3 Expectation 36
     - 2.8.4 Maximization of Expected Utility 36
   - 2.9 Where do Cardinal Utilities Come From? 38
   - 2.10 Representation Theorems 39
   - 2.11 Conclusion 41

3. Objective Bayesianism and Knowledge-First Epistemology 42
   - 3.1 Introduction 42
   - 3.2 Ur-Priors 44
   - 3.3 The Subjectivity of Subjective Bayesianism 49
   - 3.4 Further Constraints on Ur-Priors 50
   - 3.5 The Principle of Indifference 52
   - 3.6 Carnap on Confirmation 53
   - 3.7 Grue 57
   - 3.8 Whither Objective Bayesianism? 59
     - 3.8.1 Objective Bayesianism and Permissivism 59
     - 3.8.2 Is Objective Bayesianism Committed to Magic? 62
   - 3.9 Conditionalization and Ur-Prior Conditionalization 65

3.10 Knowledge-First Epistemology and E = K 67
3.11 Conclusion 68

4. Knowledge-Based Decision Theory? 70
4.1 Introduction 70
4.2 Traditional Decision Theory and Practical Dilemmas 72
4.3 The Role of Beliefs in Traditional Decision Theory 77
4.4 Knowledge-Based Decision Theory 82
4.5 Not All Knowledge Is Evidence 83
4.6 Not All Evidence Is Knowledge 86
4.7 Conclusion 92

5. Excuses, Would-Be Knowledge, and Rationality-Based Decision Theory 94
5.1 Introduction 94
5.2 Excuses and Dilemmas 95
5.2.1 Excuses in General 95
5.2.2 Williamson on Justification and Excuses 98
5.2.3 Lasonen-Aarnio on Excuses and Dispositions 100
5.3 Justification as Would-Be Knowledge 106
5.4 Conclusion: Rationality-First 115

6. Experientialism 116
6.1 Introduction 116
6.2 Three (or Four) Views 117
6.3 Against Psychologism 123
6.4 Against Factualism 128
6.5 For Experientialism 129
6.6 Inconsistent Evidence? 132
6.7 Defeaters 134
6.7.1 Pollock on Reasons and Defeaters 135
6.7.2 Defeaters for Bayesianism 139
6.8 Conclusion 146

7. The Normative Force of Unjustified Beliefs 148
7.1 Introduction 148
7.2 Broome's Interpretation of Normative Requirements 150
7.3 Chisholm's Puzzle of Contrary-to-Duty Obligations 153
7.4 A Solution to Chisholm's Puzzle 157
7.5 Normative Requirements Revisited 161
7.6 The Restrictor View of Conditionals 167
7.7 Modus Ponens and Strong Factual Detachment 169
7.8 Conclusion 172

8. The Problem of Easy Rationality 173
8.1 Introduction 173
8.2 Brief History 174
8.2.1 Fumerton and Vogel against Reliabilism 174
8.2.2 Cohen on Easy Knowledge 175

8.2.3 Pryor on Non-Quotidian Undermining 176
8.2.4 Huemer on Defeasible Justification 177
8.3 The Problem 177
8.3.1 Ampliativity 178
8.3.2 Single-Premise Closure 178
8.3.3 Mere Lemmas 180
8.3.4 Entailment 181
8.3.5 The Conflict 182
8.4 Possible Solutions 182
8.4.1 Denying Single-Premise Closure 182
8.4.2 Denying Mere Lemmas 183
8.4.3 Denying Entailment 184
8.4.4 Denying Ampliativity 185
8.4.5 The Problem of *Ex Post* Easy Rationality 186
8.5 Epistemic Principles and Evidence 187
8.6 The Problem of Easy Rationality for Experientialism 189
8.7 Conclusion 192

9. Evidentialism, Reliabilism, Evidentialist Reliabilism? 193
9.1 Introduction 193
9.2 Evidentialism 193
9.3 Reliabilism 197
9.4 How to Solve the Generality Problem 201
9.5 How to Measure Reliability 202
9.6 What is Pr? 203
9.7 Evidential Probability 207
9.8 Credences 208
9.9 Conclusion 209

10. Conclusion 211

References 213
Index 221

Acknowledgments

This book has been a long time in the making, and I am very grateful to the many friends and colleagues who have helped me along the way. Some of the material has its origins in my dissertation, written too many years ago under the direction of Ernest Sosa. Thanks, Ernie, for that, and for everything else, and many thanks also to the other members of my committee, Jamie Dreier and James van Cleve. Many people have helped me through the years by discussing with me some of the issues in this book: thanks in particular to Charity Anderson, Eduardo Barrio, Rodrigo Borges, Alex Byrne, David Christensen, Eleonora Cresto, Jonathan Drake, Janice Dowell, Julien Dutant, Andy Egan, Adam Elga, Branden Fitelson, Richard Fumerton, James Genone, Alvin Goldman, Alan Hájek, John Hawthorne, Alan Hazzlett, Thomas Kelly, Maria Lasonen-Aarnio, Keith Lehrer, Clayton Littlejohn, John MacFarlane, Kevin McCain, Alberto Moretti, Ram Neta, Eleonora Orlando, Federico Penelas, Manuel Perez Otero, Douglas Portmore, Ted Poston, Jim Pryor, Agustín Rayo, Miriam Schoenfield, Mark Schroeder, Larry Shapiro, Asaf Sharon, Alan Sidelle, Nico Silins, Declan Smithies, Elliott Sober, Levi Spectre, Scott Sturgeon, Michael Titelbaum, Jonathan Vogel, Ralph Wedgwook, Jonathan Weisberg, Roger White, and Timothy Williamson. I am grateful also to audiences at the Australian National University, the Sociedad Argentina de Análisis Filosófico, the Sociedad Española de Filosofía Analítica, the Institute Jean Nicod, the University of Iowa, the University of St Andrews, Arizona State University, University of Geneva, UNAM, the Logos group at the Universitat de Barcelona, and the Orange Beach Epistemology Conference.

Thanks also to two anonymous readers for Oxford University Press. Both of them provided wise advice and encouragement. Reader B, in particular, offered extremely detailed comments and invaluable suggestions which made the book much better. Thanks Kim Richardson for his (minimally interventionist) help with copy editing. Many thanks as well to Peter Momtchiloff for his encouragement and patience.

Some colleagues and students at the University of Arizona read a complete draft of the manuscript and discussed it with me, which resulted in many improvements: many thanks to Stew Cohen, Tim Kearl, Sarah Raskoff, Eyal Tal, Jason Turner, and Robert van't Hoff. Thanks also to the students in my Fall 2017 seminar, who endured parts of an even earlier draft.

Two philosophers who deserve special thanks are Matt McGrath and, again, Stew Cohen. With Matt we co-authored a couple of papers where we develop and defend the view that we can have false reasons, which I put to use here. Matt is an incredible philosopher and I look forward to continue working with him, for he is always very gentle with my mistakes. With Stew we also co-authored a couple of papers on knowledge-first epistemology, which I also put to use in this book. My very first published paper, on Stew's new evil demon problem, was published by him in *Philosophical Studies*, and I count myself lucky that years later I got to be his colleague and discuss that issue with him, among many others. Stew has become a big influence in my philosophical thinking in the past decade or so, and, more importantly, a dear friend.

My greatest debt by far, intellectual and otherwise, is to my wife, Carolina Sartorio. We co-authored a paper and two kids, and I'm not sure which one of those was harder. She and our sons, Tomás and Lucas, have made me happy before, during, and after writing this book. Carolina has also read the whole manuscript more than once, providing wise advice while being annoyed at the fact that there are still typos in it.

1
Introduction

Tomás wants a candy, and so grabs the candy-looking thing Lucas is offering him and puts it in his mouth. What Lucas gave Tomás was no candy, but a marble.[1] Understandably, Tomás is disappointed—but was he irrational in acting as he did? Well, it depends, of course. Is Tomás aware of his brother's playful nature? Does Tomás, in particular, have any reason to think that there is anything amiss with Lucas's offer? Let's say that he doesn't: he thinks that Lucas is genuinely being generous and sharing his Halloween bounty with him. Indeed, let us suppose that Lucas himself is unaware of the fact that there is a candy-looking marble among the candy. So, again, was Tomás irrational in acting as he did? Obviously not. And yet some of the main theories of theoretical and practical rationality on offer nowadays entail that he was. My aim in this book is to vindicate Tomás's rationality.

What was rational for Tomás to do is determined, roughly, by two things: first, what he wanted to achieve, and second, which action was rational for him to think would most likely get him what he wanted. The first factor has to do with Tomás's so-called conative attitudes—his desires, wants, intentions, etc. The second factor has to do with Tomás's cognitive attitudes—his beliefs and degrees of confidence. There is an important feature of those cognitive factors that determine what it is rational for Tomás to do: they must themselves be rational attitudes.[2] Which cognitive attitudes it is rational for Tomás to adopt is tightly related to what evidence Tomás has at the time he makes his decision. Two very influential theories of empirical evidence are Psychologism, according to which your evidence is given by the experiences you are undergoing, and Factualism, according to which your evidence is constituted by all and only those propositions you know through experience.[3] Both of these views entail that Tomás's action was irrational.

According to Psychologism, the relevant piece of evidence that Tomás has is his experience with the content that Lucas is offering him candy. Note well: Tomás's evidence, according to Psychologism, is *his experience*, not its content. His evidence, then, is compatible with the fact that what Lucas is offering him is a marble. If your evidence is compatible with the truth of a proposition, then you should consider what would happen if that proposition were true when deciding what

[1] "Se non è vero, è ben trovato"—Giordano Bruno, via my grandmother.

[2] Some philosophers think the same applies to Tomás's conative attitudes—they too must be rational. I disagree with these philosophers, but this disagreement will not matter for what follows.

[3] The term "Psychologism" is from Dancy (2000) (who opposes it). Factualism, aka E = K, is most famously defended by Williamson (2000).

Being Rational and Being Right. Juan Comesaña, Oxford University Press (2020). © Juan Comesaña.
DOI: 10.1093/oso/9780198847717.001.0001

to do. Of course, exactly what degree of confidence you should have in propositions that are not ruled out by your evidence will depend on the details of the case. Pieces of evidence can fail to rule out a proposition and yet make it highly unlikely, perhaps so unlikely as to not be worth bothering with it in the circumstances. But whether you should bother about the truth of a proposition depends not just on how confident you are of its truth-value, but also on what the consequences of acting would be if the proposition were true compared to what they would be if the proposition were false. Note also that Tomás's evidence, according to Psychologism, is compatible not only with the proposition that what Lucas is offering is not candy but a marble, but also compatible with the hypothesis that Tomás is a brain in a vat whose program is going to be terminated if he intends to eat what looks to him like candy, and, in general, with any skeptical hypothesis that entails that he is having the experiences he is having. So, in deciding what to do, Tomás should consider all of those hypotheses, according to Psychologism. But he doesn't. So, according to Psychologism, Tomás was irrational.

According to Factualism, Tomás has little evidence relevant to what to do on this occasion. He doesn't know that what Lucas is offering is a candy, because it isn't. He doesn't know either that what Lucas is offering is not a candy, because he doesn't believe this proposition. Maybe, if proponents of Factualism are lucky, Tomás knows that what Lucas is offering looks like candy. But if he knows that, he also knows that what Lucas is offering looks like a marble that looks like candy. In any case, Tomás doesn't know enough to rationalize doing what he did. So, according to Factualism, Tomás was irrational.

That consequence is unacceptable. Tomás wasn't right, but he was rational. This is how Tomás reasoned: Lucas is offering is a candy; I would like to eat a candy right now; therefore, I shall eat what Lucas is offering. (Tomás's internal monologue when deliberating is always very formal.) Notice, in particular, that Tomás took for granted that Lucas was offering candy. This is where Tomás went wrong according to both Psychologism and Factualism. I side with Tomás. In his situation, where nothing indicated that he couldn't take his experience at face value, Tomás's evidence was the content of his experience: that Lucas was offering candy. Alas, that content is false. So, there can be false evidence. That is not as bad as it sounds. This view, which I call Experientialism, is developed in the rest of the book.

Chapter 2 introduces the mathematics of probability and traditional decision theory. Readers familiar with those theories might want to skip or skim the more expository parts of that chapter.[4] I subscribe to a traditional criticism of traditional decision theory: that it has the consequence that logical omniscience is rationally required. I examine and reject some arguments against this objection—the so-called "Dutch Book" argument and the more recent accuracy-based argument—and find

[4] In writing those parts I benefited greatly from reading drafts of Titelbaum (forthcoming).

them both wanting. Nevertheless, I do adopt the framework of traditional decision theory as a useful tool to examine the issues at the core of this book. That theory is compatible with both objective and subjective views about the relation of evidential support. Chapter 3 discusses those views (under the guise of subjective and objective Bayesianism), arguing for objectivism. Chapter 3 also introduces Factualism as supplying a theory of evidence, absent in traditional decision theory.

In Chapter 4 I develop and criticize what following Dutant (forthcoming) I call "knowledge-based decision theory." Knowledge-based decision theory consists of traditional Bayesian decision theory married to Factualism as the theory of evidence. Factualism, I argue, fails in both directions: not all knowledge is evidence, and not all evidence is knowledge. My argument that not all evidence is knowledge rests on a couple of theses. First, that evidence plays a certain role in decision-making: namely, our evidence consists of those propositions that it is rational for us to take for granted in deciding what to do. Second, what I will call "Fumerton's thesis": that rational action presupposes rational belief, in the sense that it must be rational for us to believe those propositions that we take for granted in deciding what to do. Given that it is sometimes rational in deciding what to do to take for granted certain propositions which (unbeknownst to us) are false, some false propositions can be part of our evidence, and so it is rational for us to believe some false propositions.

Chapter 5 takes up a reply to my argument that there can be false rational beliefs. According to that reply, the argument fails because it doesn't consider the distinction between justification and excuses (or, depending on the author, the distinction between justification and reasonableness, or the distinction between justification and rationality, or the distinction between different kinds of justification). I examine this reply in detail, and conclude that it doesn't work. Indeed, careful reflection on the relationship between justification and dispositions to believe (which relationship is central to the reply in question) favors the existence of a notion of justification independent of knowledge. The chapter concludes by sketching this "rationality-first" conception of epistemology and decision theory.

Chapter 6 picks up the idea that there can be false rational beliefs, and more particularly the idea that we can have false evidence. This chapter defends that idea by contrasting the three conceptions of basic empirical evidence already mentioned: Psychologism, according to which our empirical evidence is constituted by our experience; Factualism, according to which our empirical evidence is constituted by what we know through experience; and Experientialism, according to which our empirical evidence is constituted by what we are basically justified in believing through experience. I end this chapter by addressing a problem for the defeasibility of empirical belief which might seem to affect Experientialism. I argue that the problem affects not just Experientialism, but Psychologism and Factualism as well, and that it has to do with the difference between how evidence justifies what it is evidence for, on the one hand, and how evidence providers (such as experience) justify belief in the evidence they provide, on the other.

Chapters 7 through 9 pick up some loose ends. As I said, in chapter 4 I defend Fumerton's thesis: that rational action presupposes rational belief. Chapter 7 then takes up the question of what we should say about the rationality of forming further attitudes once you already have an irrational belief. I connect this topic with Broome's work on normative requirements, and argue that a proper answer to the question requires us to think through Chisholm's problem of contrary-to-duty obligations. Both Chisholm's problem and the question of normative requirements can be illuminated by reflecting on dyadic deontic logic and the restrictor view of conditionals, this latter a view widely held by linguists. I end by arguing that, in deontic contexts, not only does Modus Ponens fail, but so too does "dynamic" Modus Ponens—that is to say, not only can the conclusion of a Modus Ponens argument be false when its premises are true, but the conclusion can also be false even when its premises are *known*.

Chapter 8 takes up the problem of easy rationality, which interacts in interesting ways with my argument against Factualism. My take on what the problem of easy rationality consists in is the following. Assume that inferentially ampliative knowledge is possible—that is to say, assume that it is possible to believe a proposition *H* on the basis of another proposition *E* even when *E* does not entail *H*. There is then pressure to say that anybody who knows *H* is in a position to know that *either H or not-E* (after all, *H* entails this latter proposition). But what could justify us in believing this latter proposition? *H* would seem to be the obvious answer, but given that *H* is itself justified by *E*, *H* can justify us only if *E* can. But how could *E* justify us in believing *H or not-E*, when the negation of this latter proposition (which is equivalent to *E and not-H*) entails *E*? This very general argument bears some resemblance to my argument in chapter 4 that not all knowledge is evidence. The threat, then, is that that argument from chapter 4 doesn't affect Factualism in particular, but is instead a general problem. In chapter 8 I argue that this is not so, for reasonable solutions to the problem of easy rationality leave the argument that not all knowledge is evidence untouched.

Finally, in chapter 9 I contrast the view developed in this book with both Reliabilism and Evidentialism. In previous work I argued for a combination of Evidentialism and Reliabilism, which I called "Evidentialist Reliabilism" (see, particularly, Comesaña (2010a)). Meanwhile, Richard Pettigrew (forthcoming) and Weng Hong Tang (2016b) have extended Evidentialist Reliabilism to cover the case of credences. In this chapter I take up the issue of whether the theory developed in this book can be fruitfully considered a hybrid of Evidentialism and Reliabilism. I conclude that the theory has as many differences with each of those views as it has similarities. Against Pettigrew and Tang, I argue that the probability in question at the core of Evidentialist Reliabilism has to be an evidential probability, not a physical one. In that respect, the view defended in this book differs radically from Reliabilism. On the other hand, my rejection of Psychologism in favor of Experientialism has the consequence that not all justification is justification by evidence, and to that extent the view defended here differs also from Evidentialism. Chapter 10 brings together different threads in the book and summarizes its main conclusions.

Throughout the book I speak indistinctly of rational and justified beliefs and actions. Some philosophers would balk at this, for they think that there are important differences between rationality and justification. I have not been convinced by arguments to that effect. Chapter 5 may nevertheless be relevant to this, for there I argue that distinguishing between justification and excuses will not save Factualism from my criticisms. As far as I can see, analogous arguments would apply to the distinction between justification and rationality.

2
Probability and Decision Theory

2.1 Introduction

One of the main arguments of this book, to be presented in chapters 4 through 6, starts from considerations regarding rational action and concludes with considerations regarding epistemic rationality. It is widely acknowledged that there are interesting relations between practical and epistemic rationality, and in particular that there is some kind of dependence of practical on epistemic rationality: what it is rational for us to do depends on what it is rational for us to believe. One traditional way of spelling out this dependence, the kind of traditional decision theory to be developed in this chapter, relies on the mathematical theories of probability and expected utility. I do not assume that the mathematical framework of traditional decision theory is the only possible way of discussing these issues, nor that there are no serious problems with it. Most notably, that framework obscures issues having to do with what David Lewis called "hyperintensional epistemology": the epistemology of logic and mathematics itself, for instance. But those are not the central issues to be discussed in this book, and the discipline enforced by the mathematical framework is welcome. As we shall see, there are no shortage of interesting philosophical problems that are left open by the framework. In this chapter, then, I introduce traditional decision theory, including the theory of probability. In doing so, I assume at least passing familiarity with set-theory and first-order logic.[1]

I first introduce the mathematics of probability, and then present two interpretations for the formalism: what I call "physical" and "normative" probabilities. These two kinds of probabilities will make appearances throughout the book. The claim that there are normative probabilities, in particular, is the claim that our degrees of belief (credences) must conform to the probability calculus to be rational. This thesis is sometimes called "Probabilism." On the face of it, Probabilism is obviously false. I consider two famous arguments for Probabilism, the Dutch Book argument and the Accuracy-based argument, and find that they are not sufficient to outweigh the prima-facie falsity of Probabilism. I consider also the thought that Probabilism is an idealization which may be harmless in some contexts. I agree with this idea, properly interpreted. I turn next to traditional

[1] I use the usual symbols for the connectives, quantifiers and set-theoretic operations, and I happily run roughshod over all sorts of use-mention distinctions.

Being Rational and Being Right. Juan Comesaña, Oxford University Press (2020). © Juan Comesaña.
DOI: 10.1093/oso/9780198847717.001.0001

decision theory: the theory of practical rationality as maximization of expected utility. In the context of decision theory, I discuss representation theorems as another argument for Probabilism, and find it wanting as well. Finally, I end with a statement of subjective Bayesianism, the theory according to which Probabilism and conditionalization (a claim about how credences should evolve in response to new evidence) are the only two rational requirements on credences.

2.2 Probabilities: An Axiomatic Approach

The most common approach to the theory of probability is in terms of Kolmogorov's axioms. These axioms can be seen as characterizing a function. The domain of that function can be thought of as consisting of sets or propositions. I describe both variants.

2.2.1 Sets

Although philosophers are used to thinking of probability as defined over a set of sentences or propositions, mathematicians usually think of it as defined over subsets of a set of "events," such as {1, 2, 3, 4, 5, 6} if we are thinking of the possible results of the throw of a dice, or {*Heads,Tails*} if we are thinking of the possible results of the flip of a coin. As we shall see, the two ways of formulating probability theory need not be far from each other. But Kolmogorov himself stated his axioms in a set-theoretic framework, so I start with it.

Formally, a *probability space* is a triple $<\Omega, F, P>$. Ω can be informally thought of as the set of all possible outcomes of an experiment that we are interested in. The *field* F is a set of subsets of Ω —these subsets are usually called events. Assume first that Ω is finite. In that case, F must be an *algebra* over Ω, meaning that it must include Ω itself and be closed under the operations of complementation (relative to Ω) and union—that is to say, if A and B are both in F, then so are $\bar{A}$, $\bar{B}$ (the complements of A and B) and $A \cup B$ (the union of A and B). For the finite case, we can think of F as simply the powerset of Ω—the set of all subsets of Ω. P is then a *probability function* if and only if it satisfies the following axioms:

Kolmogorov's axioms

Non-negativity: For every $A \in F, Pr(A) \geq 0$.
Normality: $P(\Omega) = 1$.
Additivity: For any two sets $A \in F$ and $B \in F$, if $A \cap B = \varnothing$, then $Pr(A \cup B) = Pr(A) + Pr(B)$.

Suppose that a set $\{A_1, A_2, \ldots, A_n\}$ *partitions* Ω, in the sense that every member of Ω is included in one and only one of the A_is. Then, by additivity, $P(A_1 \cup A_2) = P(A_1) + P(A_2)$, because none of the A_i have any members in common. But then, $A_1 \cup A_2$ and A_3 also have no members in common, and so $P(A_1 \cup A_2 \cup A_3) = P(A_1) + P(A_2) + P(A_3)$. Generalizing gives us the following theorem:

Finite Additivity: If $A_1, A_2 \ldots A_n$ is a finite sequence of pairwise disjoint sets, each belonging to F, then: $P(A_1 \cup A_2 \cup \ldots \cup A_n) = P(A_1) + P(A_2) + \ldots + P(A_n)$

Kolmogorov also definitionally introduced a two-place function over F:

Conditional Probability: $P(A|B) = \dfrac{P(A \cap B)}{P(B)}$; provided that $P(B) > 0$.[2]

Conditional probabilities thus defined are also probability functions—that is to say, for any events A, B, and C in F, $P(A|B) \geq 0, Pr(\Omega|A) = 1$, and if A and C are mutually incompatible, then $P(A \vee C|B) = P(A|B) + P(C|B)$.

Suppose now that Ω is an infinite set—for example, suppose that we are throwing darts with point-sized tips at a line fragment, and Ω is the set of all points in the line. Kolmogorov himself extended the axiomatization to this infinite case. We must now think of F as being closed under *countable* as well as under finite unions.[3] To extend the axiomatization to the countable case, we just need to replace additivity with:

Countable Additivity: If $A_1, A_2, \ldots$ is a countable sequence of pairwise disjoint sets, each belonging to F, then: $P(A_1 \cup A_2, \ldots) = P(A_1) + P(A_2) \ldots$.

Perhaps the ellipses in the formulation of countable additivity bother you. No problem! Mathematicians have devised a notation that conveniently hides the ellipses behind big symbols:

Countable Additivity: If $A_1, A_2, \ldots$ is a countable sequence of pairwise disjoint sets, each belonging to F, then:

$$P(\bigcup_{i=1}^{\infty} A_i) = \sum_{i=1}^{\infty} P(A_i)$$

[2] Given that the functions take different kinds of things as arguments, we can abuse notation by calling them both P.

[3] This makes F a so-called σ-algebra.

An important issue that arises in the infinite case is that, sometimes, F will have to be a proper subset of the powerset of Ω. This is because not all subsets of Ω will be "well-behaved" enough to receive a probability. I come back to this issue in section 2.5 and in the discussion of the regularity constraint in chapter 4.

2.2.2 Propositions—and Bayes's Theorem

Most philosophers will be familiar with probability as applied to propositions, not sets. An axiomatization over sets and an axiomatization over propositions need not be as far from each other as it first appears. For we can think of Ω as a set of possible worlds (perhaps all of them, perhaps only a subset we are interested in). Some philosophers think of propositions as being sets of possible worlds.[4] If so, then the two axiomatizations are equivalent when Ω is a set of possible worlds. And even when Ω is a set of possible outcomes, we can reconceptualize those possible outcomes as equivalence classes of possible worlds—for instance, 1 is the equivalence class of all the possible worlds where the die comes up 1. Some other philosophers, however (perhaps the majority) do not simply identify propositions with sets of possible worlds. But we can still think of propositions as determining (although perhaps not being determined by) sets of possible worlds. For better or worse, it is a consequence of the probability axioms that if two propositions are logically equivalent then they receive the same probability. That means that the probability function itself is not concerned with the possible differences between propositions that are true in all the same possible worlds.

Assume that we have a set of propositions S which is closed under negation and disjunction. That is to say, if ϕ and ψ are in S, then so are $\neg\phi$ and $\phi \vee \psi$.[5] A function Pr is a probability function defined over S if and only if it satisfies the following axioms:

Kolmogorov axioms

Non-negativity: For all propositions $\phi \in S, Pr(\phi) \geq 0$.
Normality: For any tautology $T \in S, Pr(T) = 1$.
Additivity: For any two mutually incompatible propositions $\phi \in S$ and $\psi \in S, Pr(\phi \vee \psi) = Pr(\phi) + Pr(\psi)$.

Together, non-negativity and normality constrain the range of any probability function to the real numbers between 0 and 1. To some extent, these first two axioms have a book-keeping nature, for the top and bottom values are arbitrarily

[4] Most prominently Stalnaker and Lewis—see, for instance, Stalnaker (1984) and Lewis (1986).

[5] Notice that the pair of negation and disjunction is complete in the sense that we can represent every truth-function with them.

chosen. However, they also have decidedly substantive consequences, for (together with additivity) they entail that probabilities have both a minimum and a maximum value, and that all tautologies receive the maximum value.

Suppose that the set of sentences $\{\phi_1, \ldots, \phi_n\} \subseteq S$ *partitions* S, i.e., exactly one of the ϕ_is must be true. Then, by additivity, $Pr(\phi_1 \vee \phi_2) = Pr(\phi_1) + Pr(\phi_2)$, because ϕ_1 and ϕ_2 are mutually incompatible. But $\phi_1 \vee \phi_2$ is, in its turn, mutually incompatible with ϕ_3. So $Pr(\phi_1 \vee \phi_2 \vee \phi_3) = Pr(\phi_1) + Pr(\phi_2) + Pr(\phi_3)$. Generalizing, if $\{\phi_1, \ldots, \phi_n\} \subseteq S$ *partitions* S, then:

$$Pr(\phi_1 \vee \ldots \vee \phi_n) = Pr(\phi_1) + \ldots + Pr(\phi_n)$$

which is finite additivity. The generalization to the countable case gives us countable additivity:

Countable Additivity

$$Pr(\bigvee_{i=1}^{\infty} \varphi_i) = \sum_{i=1}^{\infty} Pr(\varphi_i)$$

Besides the one-place unconditional probability function *Pr*, a different two-place conditional probability function is often definitionally introduced as follows:

Conditional Probability

$$Pr(\phi|\psi) = \frac{Pr(\phi \wedge \psi)}{Pr(\psi)}; \text{ provided that } Pr(\psi) > 0.$$

Conditional probabilities thus defined are also probability functions—that is to say, for any propositions ϕ, ψ, and χ and any tautology T, $Pr(\phi|\psi) \geq 0$, $Pr(T|\psi) = 1$, and if ϕ and χ are mutually incompatible, then $Pr(\phi \vee \chi|\psi) = Pr(\phi|\psi) + Pr(\chi|\psi)$. At the extremes, when $\psi \supset \phi$ is a tautology, $Pr(\phi|\psi) = 1$, and when $\psi \supset \neg\phi$ is a tautology, $Pr(\phi|\psi) = 0$ (provided that $Pr(\psi) \neq 0$).

Notice that, given this definition, we can carry out the following proof:

1. $Pr(\phi|\psi) = \dfrac{Pr(\phi \wedge \psi)}{Pr(\psi)}$

2. $Pr(\psi|\phi) = \dfrac{Pr(\phi \wedge \psi)}{Pr(\phi)}$

3. $Pr(\psi|\phi)Pr(\phi) = Pr(\phi \wedge \psi)$

4. $Pr(\phi|\psi) = \dfrac{Pr(\psi|\phi)Pr(\phi)}{Pr(\psi)}$

Steps 1 and 2 are applications of the definition of conditional probability, with step 2 using also the commutativity of conjunction. Step 3 is the result of multiplying both sides of 2 by $Pr(\phi)$. Finally, step 4 substitutes $Pr(\phi \wedge \psi)$ in 1 with $Pr(\psi|\phi)Pr(\phi)$—and those quantities are identical to each other according to 3. Step 4 is also known as Bayes's theorem, which relates the conditional probability $Pr(\phi|\psi)$ to the converse conditional probability $Pr(\psi|\phi)$ (also known as the "likelihood" of ψ given ϕ), and is central to the interpretation of probability as a measure of subjective uncertainty—on which more in section 2.7.

This axiomatization is not the only one available. Most notably, there are alternative axiomatizations that take the two-place conditional probability function as primitive, and then define the unconditional probability of a sentence ϕ as follows: $Pr(\phi) = Pr(\phi|T)$.[6]

2.3 Probabilities: A Measure-Theoretic Approach

Besides the axiomatic characterization of the probability function, we can also think of probability as a branch of measure theory. A *measure* is a function defined over a set Ω which assigns a non-negative real number to each suitable subset of Ω.[7] Informally, a measure gives the *size* of a subset. A measure, moreover, must be additive—that is to say, it must obey the following constraint:

Additivity: If S and S' are suitable subsets of Ω and $S \cap S' = \varnothing$, then $M(S \cup S') = M(S) + M(S')$.

Thus, for example, take Ω to be the set of all the states in the United States of America. The function that assigns to each state its population is then a measure, for it assigns a non-negative number to each state, and the number that it assigns to any subset of states (for instance, New England) is the sum of the number that it assigns to each of its member states.

Notice that, given additivity, no subset of Ω will have a measure greater than the measure assigned to Ω itself. Given any measure M we can define another measure M' which is the result of normalizing M to 1 (or, indeed, to any number we want). The way to do that is to let $M'(\Omega) = 1$ and $M'(S) = \frac{M(S)}{M(\Omega)}$ for any suitable $S \in \Omega$. If we normalize the measure that measures the population by state, then the result is a measure of the proportion that each state contributes to the total population.

[6] See, for instance, Renyi (1970) and Popper (1959).

[7] "Suitable" because if Ω is large enough then some of its subsets won't be measurable. See section 2.5 and chapter 4 for a discussion of the regularity constraint.

Suppose now that Ω is a set of possibilities—perhaps possible worlds. Any normalized measure over Ω will now count as a probability function. Thus, we can think of probabilities as *normalized measures over possibility space.*

2.4 Probabilities: Stochastic Truth-Tables

Thinking of probabilities as normalized measures over possibility space affords us yet a different way of thinking about them, which traces back at least to Carnap.[8] Consider the familiar truth-table for two propositions ϕ and ψ:

ϕ	ψ
T	T
F	T
T	F
F	F

That truth-table can also be seen as representing the set Ω of all possible combinations of truth-values between ϕ and ψ. Carnap called each one of those combinations a "state description." Consider then a normalized measure over Ω—i.e., consider a probability function defined over Ω. A stochastic truth-table is then just a regular truth-table with an additional column that represents the probability measure defined over that possibility space:

ϕ	ψ	Pr
T	T	a
F	T	b
T	F	c
F	F	d

The only constraint is that each of a, b, c, and d have to be non-negative, and they have to add up to 1. Once we conceive of probabilities as representable by stochastic truth-tables, we can also conceive of probability calculations as algebraic manipulations of the assigned probabilities. For instance, $Pr(\phi \supset \psi) = a + b + d$, whereas the conditional probability of ϕ given ψ, $Pr(\phi|\psi) = \frac{Pr(\phi \wedge \psi)}{Pr(\psi)}$, is $\frac{a}{a+b}$. Notice that any proposition that is logically equivalent to a truth-functional combination of

[8] See Carnap (1950), and also Fitelson (2008).

ϕ and ψ will be logically equivalent to a disjunction of state-descriptions. Thus, the proposition ϕ is logically equivalent to the disjunction of the first and third state-descriptions on the table. Given that *Pr* is a measure, the probability that *Pr* assigns to a proposition will be simply the sum of the probabilities that it assigns to each state-description where it obtains.

2.5 Problems with Conditional Probability

Remember that a conditional probability $Pr(\phi|\psi)$ is defined only when $Pr(\psi)$ exists and is positive. So, if $Pr(\psi) = 0$ or is undefined, then $Pr(\phi|\psi)$ is undefined. This leads to problems.[9]

Consider the cases where a probability function is defined over an infinite set—for instance, a set with the cardinality of the continuum. This is not an idle possibility, as physics seems to deal, at least on the face of it, with continuous functions all the time, and we may want to say that there are physical probabilities (see section 2.6). To dramatize the situation, suppose that you are throwing darts at a target, and that the darts' tips are point-sized. Suppose that there is an area of the target where the dart has an equal chance of falling on any of the points in that area. It follows immediately from countable additivity that that chance has to be 0. For if it is a positive number, then the probability that the dart hits the area is greater than 1. That doesn't mean that we cannot assign meaningful probabilities over that area—one possibility is to assign probabilities to sets of points in a manner proportional to the area occupied by that set.[10] Thus, we can assign some probability to your hitting a point in the whole area, and half that probability to your hitting a point in the western hemisphere of the area (assuming that the area is circular). That means, however, that any subset of points which corresponds to an area of dimension 0 will have to be assigned a probability 0—for instance, the probability that the dart hits anywhere in a chord has to be 0. That already signals something weird about probabilities distributed over an infinite set of events, for surely it is not impossible for you to hit a point in a particular chord—after all, every point is in some chord! Worse yet, suppose that we ask what the probability is that the dart hit the right-hand side of the circle, given that it hit the central diameter. Surely this probability should be ½. And yet, the probability that you hit the right-hand side of the circle given that you hit the central diameter is undefined, because the probability that you hit the central diameter is 0.

[9] What follows is taken from Hájek (2003) and Hájek (2011), who argues on the basis of these and other problems against what he calls the "ratio analysis" of conditional probability—"analysis," and not "definition" because Hájek argues that the counterexamples show Kolmogorov's definition is not true to an intuitive notion of conditional probability.

[10] This corresponds to a Lebesgue measure over the area.

Some authors (see, for example, Lewis (1980)) have suggested that we solve the problem of division by 0 by appealing to "infinitesimals"—numbers greater than 0 but smaller than any real number. But even if we accept this suggestion there will be some events that cannot be assigned any probability at all—these correspond to unmeasurable subsets of the area of interest. And yet, surely the probability that you hit a point in, say, a Vitali set given that you hit a point in that very Vitali set has to be 1, not undefined.[11]

Hájek (2003) argues, on the basis of these problems, for taking conditional probability as primitive, which allows for meaningful probabilities in the cases considered. As we saw before, it is possible to axiomatize the notion of conditional probability directly, and then define unconditional probability in terms of conditional probability.

2.6 Physical Probabilities

So far we have presented different ways of looking at probabilities considered as mathematical objects. But the philosophical interest in probability gets going when we ask whether there is anything that answers to the formalism. This question falls under the traditional heading of "interpretations of probability," but I won't here rehearse the usual suspects (different forms of frequentism, propensity theory, the classical interpretation, the logical interpretation, etc.) except when they are relevant to my concerns. Instead, I want to consider two large *kinds* of interpretations which will have a bearing in the discussion of the chapters to follow.

There is no good name for the first kind of interpretation of probability that I want to consider, but I will call them "physical probabilities." That is not a good name because I want to include the kind of probabilities that sciences other than physics, like biology or the social sciences, deal with. But "objective probabilities" is an even worse name, for the other kind of interpretation takes probabilities to be

[11] Take the interval $[0, 1)$ and suppose that we bend it into a circle. Start by first defining an equivalence relation $\sim$ on the interval such that $x \sim y$ if and only if $x - y$ is a rational number. Given the axiom of choice, we can form a set containing exactly one representative from each of the cells in the equivalence relation. That is a Vitali set. That set can be used to generate countably many "clones" of itself by picking a different representative from each cell, while taking care that each new representative is the same distance from the corresponding old one. These countably many clones partition $[0, 1)$. Given a uniform probability distribution over $[0, 1)$, no Vitali set can be assigned any probability. Given uniformity and the fact that all the sets are clones of each other, if they have any probability then they have all the same probability. Now, given Countable Additivity, the probability assigned to the whole interval (namely, 1) is the sum of the probabilities assigned to the all the clones. This probability cannot be 0—otherwise the probability of the whole interval would also be 0. The probability cannot be any positive real number, because then the probability of the whole interval would be greater than 1. Finally, the probability cannot be an infinitesimal either, because a countable sum of infinitesimals adds up to less than 1. See Rayo (2019), chapter 6, for a detailed explanation of this result.

just as objective as this kind. Enough complaining about the names. What I mean by physical probabilities is the probabilities that a complete science would assign to different events. Thus, suppose that (as some interpretations of quantum mechanics will have it) the fundamental laws of the tiny physical bits of our universe are probabilistic in nature: that is to say, given a complete description of the world at the micro level at a time t, the true and complete physical laws do not determine, together with this description, a unique complete description of the world at the micro level at a later time t'. Rather, what they do is assign certain probabilities to more than one complete description at the micro level at t'. The physical interpretation of probability holds that we should take that "probability" talk in the description of indeterministic laws literally: we can think of the laws of an indeterministic physics as a conditional probability function which states the probability of one state of the world on another.[12]

Whether they are micro or macro, physical probabilities clearly *evolve*. It may be that, at noon, the physical probability that this atom will decay at 12:01 p.m. is, say, ⅓, but by 12:02 p.m. it is certainly either 1 (if it did decay) or 0 (if it did not). How can we account for this evolution of physical probabilities? A natural answer is that physical probabilities evolve by Conditionalization on the facts. Say that a probability function Pr' results by *conditionalizing* another probability function Pr on ψ if and only if, for any proposition ϕ:

Conditionalization: $Pr'(\phi) = Pr(\phi|\psi)$

Thus, if Pr is the microphysical probability function at noon and Pr' is the microphysical probability function at 12:02 p.m., then Pr' results from conditionalizing Pr on the conjunction of all the microphysical truths at 12:02 p.m. Thus, if the atom in question did decay at 12:01 p.m. (call this proposition D), then $Pr'(D) = 1$, and if it didn't decay $Pr'(D) = 0$. It is common to call Pr the *prior* probability function, and Pr' the *posterior* probability function, when the probabilities are considered from the afternoon point of view.

In fact, we may think of *the* (say, micro) physical probability function as a kind of *ur-prior*, Pr_u, which encodes the evolution of physical probabilities in its assignment of conditional probabilities. At any given time t, the function which describes the physical probabilities at t will be a Conditionalization of Pr_u on all the facts that have obtained up to t. Notice that if a probability function Pr results by Conditionalization on a proposition ψ, then $Pr(\psi) = 1$, because, for any Pr' and any ψ, $Pr'(\psi|\psi) = 1$. Moreover, Conditionalization preserves extreme probability assignments—in a series of probability functions $Pr_1, \ldots, Pr_n$ such that Pr_i

[12] Elliott Sober (2010) has argued that we should take seriously physical macro probabilities too, for instance those appealed to in biology and other special sciences, and presumably in ordinary explanatory practices as well.

results from conditionalizing Pr_{i-1} on a proposition, if a proposition ψ is conditionalized on by Pr_i, then, for every $j \geq i$, $Pr_j(\psi) = 1$. This is not a problem for the physical interpretation of probability, for once a fact obtains it stays obtained, but it is a result worth remembering.

What *is* a problem for the physical interpretation of probability and the thesis that such probabilities evolve by Conditionalization is the set of considerations against the definition of conditional probability as a correct explanation of the intuitive notion rehearsed in the previous section. For if physical probabilities are defined over a set of continuum-many events, then there will be events that can only be assigned probability 0 and events that can be assigned no probability at all, but which are nonetheless physically possible, and thus might obtain, and thus the physical probability function should conditionalize on them. The alternative formalizations where conditional probability is taken as a primitive, mentioned earlier, might provide some relief here.

2.7 Normative Probabilities: Probabilism and Conditionalization as Necessary Conditions on Rational Credence

I will call the second kind of interpretation of probability "normative probabilities." To understand the normative interpretation of probabilities, we have to first consider the issue of which doxastic attitudes there are. Traditional epistemology is also sometimes called "binary" epistemology because it recognizes only an all-or-nothing attitude of belief. It is true that many epistemologists have also recognized the importance of taking suspension of judgment as a bona fide doxastic attitude alongside belief and disbelief (which may perhaps be identified with belief in the negation of the corresponding proposition). But what characterizes this kind of traditional epistemology is the lack of theorizing about graded doxastic attitudes. Traditional epistemologists, however, do think of justification, or rationality, as coming in degrees. Besides the issue of how much justification is needed for a belief to be outright justified (which gives rise to the different versions of contextualism, subject-sensitive invariantism, and relativism in the literature),[13] this graded notion at the heart of traditional epistemology has not been the focus of much attention—for instance, not much work has been done on whether it can itself be understood in probabilistic terms.[14]

The doxastic states countenanced by traditional epistemology figure prominently as one half of the states used in our everyday explanation and prediction of

[13] On contextualism, see for instance Cohen (1988); on subject-sensitive invariantism, see Fantl and McGrath (2002) (the name is from Hawthorne (2004a)); and on relativism see MacFarlane (2010).

[14] But see Shogenji (2012) and Moss (2018).

behavior. You went to the movies, for example, because you wanted to meet your friend, and you believed that your friend was at the movies. This explanation of your behavior appeals to what you want and what you think. The explanation tacitly assumes your rationality—even if you had the same pair of wants and thoughts, you may have irrationally gone to the park.

The case is too simple to capture most of what makes decision-making interesting. Let's complicate it a bit. Let's suppose that you cannot communicate now with your friend, so you have to rely on what he told you yesterday. He told you that he would be at the movies today if it rained, and at the park otherwise. You want to meet your friend, but you would really hate to be at the movies alone. Being at the park alone, while not as good as being with your friend, is not so bad. Also, the weather report predicts a 75 percent chance of rain for today. What should you do?

A complete answer to that question will have to await to later sections, and it will depend on details we have not yet explored. But let us concentrate here on one of those details. In the case as described, do you believe that it will rain today? We stipulated that you know that the weather report predicts a 75 percent chance of rain. Perhaps you also heard that TV reporters usually exaggerate the chance of rain. Given that this is all the information you have about whether it will rain today, do you believe that it will rain? Notice that, if you do, then the decision problem has a trivial solution: you should go to the movies, because, given your other belief that your friend will be there if it rains, it will result in your meeting your friend, which is your most preferred outcome.[15] If you don't believe it, then maybe you believe that it will not rain. In that case, again, the decision problem has a trivial solution: you should go to the park, because, given your other belief that your friend will be there if it doesn't rain, it will result in your meeting your friend, which is, again, your most preferred outcome. Or perhaps you suspend judgment on whether it will rain. In that case, the decision problem no longer has a trivial solution—but that victory is Pyrrhic, for it seems to have no solution at all. If all we can say about your opinion regarding whether it will rain is that you suspend judgment, that doesn't distinguish between going to the park and going to the movies. And it's not as if the upshot is that both actions are equally rational: we simply have not been told enough about your state of mind to figure out which action is rational.

2.7.1 Decision under Uncertainty?

There is a sub-field of decision theory, known as "decision under uncertainty" or "decision under ignorance" (and contrasted with "decision under risk," which

[15] Of course, it may be that your friend is not actually at the movies, even if it rains. This variation on the case forms the core of one of my arguments in chapter 4.

usually denotes the kind of traditional decision theory explained below in section 2.8), which aims to develop principles of rational decision for precisely these sorts of cases. But the principles of decision under uncertainty either do not apply to some interesting cases or do not have normative force.

A principle that does have real normative force is the *Dominance* principle.[16] Suppose that the subject can rank all the possible outcomes of all of her available actions in terms of which ones she prefers to which other ones. Then, if for a given action a_i there is another action a_j such that the subject prefers every outcome of a_j to every corresponding outcome of a_i, we say that a_i is *dominated by* a_j (more on outcomes and actions in section 2.8). Dominance then says that a dominated action is irrational. This principle does have normative force because if you choose a dominated action you could have done better (by your own lights) no matter what. But Dominance does not apply to all cases, particularly to the interesting cases. In our example, neither action dominates the other (going to the movies has a better outcome than going to the park if it rains, but going to the park has a better outcome than going to the movies if it doesn't rain).

Other principles of decision under uncertainty are intended to apply to every case, not just to those where some actions are dominated. The *Maximin* principle, for instance, instructs subjects to choose so as to maximize the minimal value obtainable with an act. Maximin would tell you, for instance, to go to the park, for, as stipulated, the worst possible outcome of going to the park (being alone at the park) is better than the worst possible outcome of going to the movies (being alone at the movies). Maximin does apply generally (or, at least, to those cases where the subject's preferences order all the possible outcomes), but its normative force is highly suspect. For why are we rationally compelled to choose as if the worst possible outcome were to occur? Why couldn't we choose as if the best possible outcome were to occur? Indeed, this is what the *Maximax* principle would have us do *always*. The Maximax and the Maximin principle are incompatible with each other, which is an indication that neither of them can be right. Neither extreme optimism nor extreme pessimism (or, to put it in decision-theoretic jargon, neither extreme risk aversion nor extreme risk seeking) is rationally mandatory.

So, if we stay within a traditional, binary conception of our doxastic attitudes, then the decision problem I sketched (and many like it) will be either trivial or impossible. But it certainly doesn't feel that way—some decision problems seem to have a non-trivial solution. The one we sketched might be one of them. How could that be? The obvious answer is that our doxastic attitudes include more fine-grained attitudes than mere belief, disbelief, and suspension of judgment. These degrees of belief, or "credences," may have the necessary structure to render at

[16] I am assuming here that you take the states to be independent of your actions. For those cases where you don't, either the Dominance principle will have to be restricted (if you take the right view about those cases) or large bullets will have to be bitten.

least some decision problems interesting. Thus, suppose that your credence in the proposition that it will rain (measured in the 0 to 1 interval) is 0.65 (remember that you know that TV weather reports tend to exaggerate the probability of rain). In that case, there will be a non-trivial answer to the question about what you should do. What that answer is will depend on the exact nature of your preferences for the four possible outcomes of being either at the movies or at the park alone or accompanied by your friend. Later sections describe traditional decision theory, which spells out in detail the nature of this dependence. But first I examine in more detail normative aspects of credences.

2.7.2 Probabilism and the Problem of Logical Omniscience

The normative interpretation of probability has it that to say that the probability for S of p is x is to say that the credence that it is rational for S to assign to p is x. The fundamental principle of the normative interpretation is the following:

Probabilism: The credences of a rational agent obey the probability calculus.

According to Probabilism, the following are requirements of rationality: your credence in any proposition must lie in the [0,1] interval of real numbers, tautologies should receive credence 1, and your credence in disjunctions of mutually exclusive propositions must be the sum of your credences in their disjuncts.

On the face of it, Probabilism is clearly false. Given that if p entails q the probability of p cannot be greater than the probability of q, Probabilism has the consequence that no rational subject can assign to p a greater credence than to q. Moreover, given Normality, Probabilism has the consequence that no rational subject can assign credence less than 1 to any tautology. Both of these consequences are counterexamples to Probabilism: it is no rational requirement to know, of any tautology, that it is a tautology, nor is it a rational shortcoming to not be aware of the fact that one proposition entails another.

In the next two sections I discuss two prominent arguments for Probabilism—the Dutch Book argument and the accuracy-based argument. At the heart of both arguments lie two mathematical theorems. The theorems as such are, of course, unimpeachable, but their philosophical interpretation—namely, that they support Probabilism—is highly suspect. The Dutch Book argument and the accuracy-based argument are (ineffective) arguments for what can be called *strong* Probabilism: the thesis that credences must be probabilistically coherent. In section 2.7.2.3 I consider, and reject, a weaker version of Probabilism: one that says that credence functions *known* to be probabilistically incoherent are irrational. In section 2.7.2.4 I consider some maneuvers intended to *defend* Probabilism from the logical omniscience problem. I endorse a kind of defense of Probabilism which does not assume that logical omniscience is the truth about ideal agents, but rather a sometimes harmless falsehood about real agents.

2.7.2.1 Dutch Book Arguments for Probabilism

To understand Dutch Book arguments for Probabilism, we need to think of bets on the truth-value of propositions.[17] In particular, we are interested in bets that pay \$1 if a certain proposition is true, and nothing otherwise. For instance, consider the bet that pays \$1 if it rains tomorrow, and nothing otherwise. How much are you willing to pay for that bet? That is to say, how much will you pay a bookie for him to pay you \$1 if it rains tomorrow, and nothing otherwise? Call the maximum that you are willing to pay for a bet of that kind your "betting price." We will also assume that your betting price is your minimum selling price—that is to say, if you are willing to pay up to \$$n$ for a bet on a proposition P (which, remember, pays \$1 if P is true and nothing otherwise), you are also willing to undertake the obligation to pay \$1 if P is true in exchange for \$$n$.

The theorem behind the Dutch Book argument says that if your betting prices are not probabilistically coherent, then there is a series of bets which you are committed to making which guarantees that you will lose money. Importantly, if your betting prices do not violate Probabilism, no such Dutch Book can be made against you.

How can that theorem be turned into an argument for Probabilism? Early probabilists, such as Ramsey (1926), simply identified your credences with your betting prices—your credence that it will rain tomorrow just *is*, for them, your maximum betting price on the proposition that it will rain tomorrow. Given that crude behaviorism, the theorem implies that your credences are not probabilistically coherent if and only if you are committed to a series of bets that guarantees a financial loss. We have to also assume that making yourself vulnerable to such sure (and avoidable) loss is irrational.

The transformation of the theorem into an argument then relies on two assumptions: that your credences are identical to your betting prices and that the prospect of a sure financial loss is irrational. A big problem is that both assumptions are false.

First, your credences need not match your betting prices—perhaps you don't like betting, so you simply don't have betting prices, or perhaps you love betting so much so that your betting prices are way above your credences.

Second, even granting that your credences match your betting prices, susceptibility to a Dutch Book may be a symptom of *practical* irrationality (although even that much is debatable—there aren't, after all, any clever bookies that are

[17] For more on Dutch Book arguments, see Hájek (2008b).

looking to profit from your Dutch Book-ability), but how does this get transmuted into an *epistemic* failure?

In answer to both of these questions, some philosophers (like Armendt (1993)) argue that susceptibility to a Dutch Book simply dramatizes a kind of inconsistency, in that we are evaluating the same thing in different ways, depending on how it is described. But take a very complicated tautology, one which, pre-theoretically, we would never say it is a rational requirement to assign credence 1 to. Let's call that tautology T^*, and let's see how the strategy under consideration works regarding this simple example.

Armendt himself treats that case merely in a footnote, presumably because he thinks it is obvious. We are interested now in a bet which pays \$1 if T^* is true and nothing otherwise. Let us assume that your credence in T^* is less than 1, say 0.7.[18] That means that you are committed to thinking of the following exchange as fair: I give you \$.7, and you give me \$1 if T^* is true, and nothing otherwise—that is to say, you are committed to thinking that selling me a bet on T^* at \$.7 is a fair deal. (How do we figure out whether T^* is true? Good question. Perhaps an oracle tells us.) But, of course, T^* is guaranteed to be true, and so you will have to pay me \$1 no matter what. That is to say, the exchange that you are committed to thinking is fair is equivalent to the following exchange: I give you \$.7 and you give me \$1. But you should not consider that deal as fair at all. Therefore, if your credences violate Probabilism, you are committed to treating equivalent deals differently. This is what Armendt describes as "divided mind" inconsistency.

But this couldn't possibly be an answer to the worry we started out with. That worry, remember, is that T^* is so complicated that we cannot determine that it is a tautology—and that this means that it is no requirement of rationality that we assign T^* credence 1. In fact, the contrary seems to be a requirement of rationality: we should refrain from assigning T^* credence 1. In defense of normality, we are now told that to do that—to assign T^* credence < 1—is to exhibit "divided mind inconsistency." But this alleged defect consists in nothing more and nothing less than having different attitudes towards lotteries involving T^* than towards gifts of money. A lottery involving T^* is, of course, a gift of money—but the whole point of the example is that we can be rationally unaware of that. So-called divided-mind inconsistency, in this case, is nothing more than another name for rational ignorance of logical facts.[19]

2.7.2.2 Accuracy-Based Arguments for Probabilism

Dutch Book arguments for Probabilism do not have the same luster they used to—there seems to be some significant consensus that they do not establish

[18] If it is more than 1, then we simply make you buy the bet instead of selling it.

[19] Analogous objections can be leveled against Christensen's "depragmatized" versions of Dutch Book arguments—see, for instance, Christensen (1996).

Probabilism. An argument originating with Joyce (1998) has taken its place as the go-to argument for Probabilism.[20] The argument is decision-theoretic in form, and it takes some stage-setting to present it in a way that does it justice. However, I do not think that it fares any better than the Dutch Book argument as a defense of Probabilism.

The philosophical motivation that at least some proponents of the accuracy-based argument (including Joyce himself) have is based on the vague but persistent thought that epistemic justification must have something to do with truth. In the realm of binary epistemology, Reliabilism is thought to be justified by some such thought, but Cohen's "new evil demon" problem (Cohen (1984)) is a resilient thorn on the side of that argument.[21] From the point of view of someone who takes the new evil demon problem seriously (not everyone does, of course), the idea that epistemic justification must have something to do with truth cannot be thought to straightforwardly support any substantive epistemological theory. Proponents of accuracy-based arguments for Probabilism have not, I think, seriously considered this dialectic—for I do not think that accuracy-based arguments can do more for Probabilism than the vague thought about the relationship between epistemic justification and truth can do for Reliabilism.

The main technical term used in the argument is that of the *accuracy* of a credence function. The guiding idea here is that accuracy is to credences what truth is to beliefs. Of course, truth is binary, but accuracy is gradational. If a proposition is true, then the higher the credence in it the more accurate it is, whereas if a proposition is false, the lower the credence the more accurate it is.

How do we measure the accuracy of a credence function? In fact, in line with many authors, let us think of the *in*accuracy of credence functions—when it comes to inaccuracy, of course, smaller numbers are better. Let us first figure out how to measure the inaccuracy of the credence that the function assigns to a single proposition P. Let that credence be n. Perhaps the most straightforward way to think of the inaccuracy of a credence assignment is by taking the absolute difference between n and a number that encodes the truth-value for P, where 1 is the code for truth and 0 is the code for falsehood. In that case, if P is true, then the inaccuracy of assigning n to P will be $1 - n$ and if P is false the inaccuracy will be n. But, for various technical reasons, the measure of distance from the truth that is usually used is not the absolute difference but the square of the difference. So, if P is true the inaccuracy of the credence assignment is $(1-n)^2$, and if P is false it is n^2. That is the inaccuracy of a single credence assignment. We can generalize this into the inaccuracy of a whole credence function by summing over the inaccuracies of

[20] For a book-long development of this kind of accuracy-first epistemology, see Pettigrew (2016).

[21] I myself have had my go at defending Reliabilism from the new evil demon problem in Comesaña (2002). I take up this and related issues regarding Reliabilism in chapters 5 and 9.

each credence assignment. Thus, the local inaccuracy of an assignment of n to proposition P relative to a possible world w is given by the following formula:

$$LI_P(n,\ w) = \begin{cases} (1-n)^2 \ \textit{if P is true in w} \\ n^2 \ \textit{if P is false in w} \end{cases}$$

and the global inaccuracy of a credence function Cr relative to a world w is:

$$GI(Cr, w) = \sum_{P_i} LI_{P_i}(Cr(P_i), w)$$

where the P_i are the propositions over which Cr is defined. That is to say, relative to a world w, the inaccuracy of a credence function is simply the sum of the inaccuracies of the credences it assigns to each proposition over which it is defined, and the inaccuracy of a credence assignment to a single proposition P is the square of the difference between that assignment and 1 if P is true, and just the square of that assignment if P is false. The actual inaccuracy of a credence function is given by applying the above formula with the actual world as w.

But, of course, to know the actual inaccuracy of a credence function we would have to know which propositions are true—and if we knew that, why bother with this whole inaccuracy business and not just fully believe all the true propositions (and fully disbelieve all the false propositions) instead? Still, we can compare the inaccuracy of two credence functions relative to *every* possible world. If one of them has lower inaccuracy than the other one relative to *each* possible world, then we say that second one is *accuracy-dominated* by the first one. The mathematical theorem behind accuracy-based arguments for Probabilism is the following: every probabilistically incoherent credence function is accuracy-dominated by a probabilistically coherent credence function.

Let me illustrate this theorem with an example of two credence functions (one probabilistically incoherent, the other probabilistically coherent) defined over just one proposition P and its negation $\neg P$. Let the incoherent function, Cr_i, assign 0.7 to P and 0.5 to $\neg P$. There are just two possible worlds to consider: w_1, where P is true and $\neg P$ is false, and w_2, where P is false and $\neg P$ is true. The global inaccuracy of Cr_i relative to w_1 is $GI(Cr_i, w_1) = LI_P(.7, w_1) + LI_{\neg P}(.5, w_1) = .09 + .25 = .34$; and the global inaccuracy of Cr_i relative to w_2 is $GI(Cr_i, w_1) = LI_P(.7, w_2) + LI_{\neg P}(.5, w_2) = .49 + .25 = .74$. Meanwhile, consider the probabilistically coherent credence function Cr_c, which assigns 0.6 to P and 0.4 to $\neg P$. The global inaccuracy of Cr_c relative to w_1 is $GI(Cr_c, w_1) = LI_P(.6, w_1) + LI_{\neg P}(.4, w_1) = .16 + .16 = .32$; and the global inaccuracy of Cr_c relative to w_2 is $GI(Cr_c, w_1) = LI_P(.6, w_2) + LI_{\neg P}(.4, w_2) = .36 + .36 = .72$. As promised, for each $w_i, GI(Cr_c, w_i) < GI(Cr_i, w_i)$—that is to say, Cr_i is accuracy-dominated by Cr_c

That was not at all a proof of the theorem, but simply an illustration of an instance of it. But the dominating coherent credence function was not chosen randomly, as can be seen by considering the following graph:

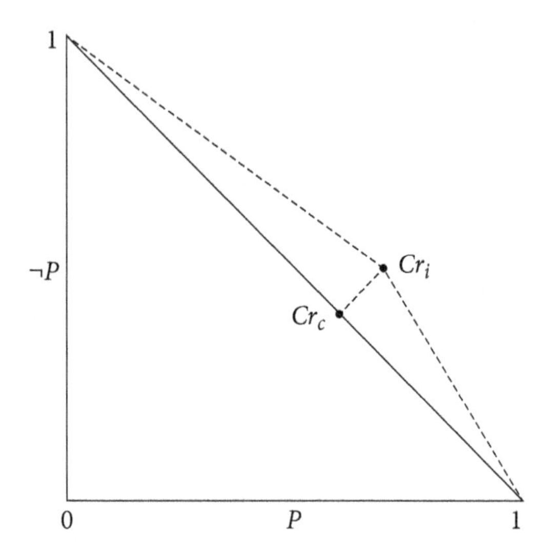

In that graph, the vertices (1,0) and (0,1) represent w_1 and w_2 respectively. The diagonal line connecting those two vertices represents all the probabilistically coherent credence assignments to P and $\neg P$. That line contains all the *(x,y)* points that add up to 1—as probabilistic coherence requires. Also plotted is the point (.7, .5), which are the credence assignments of Cr_i. And notice that Cr_c, the point (.6, .4), is the projection of the Cr_i point onto the line that contains all the probabilistically coherent credence assignments. It can easily be seen that, for any point outside that line, the projection of it on the line will be closer to *both* w_1 and w_2.[22] To see why the theorem holds, all we have to do is to generalize this two-dimensional case to arbitrary dimensions (easy, right?).

How can this theorem be turned into an argument for Probabilism? The most direct way to do it would be to hold that credence functions which are accuracy-dominated (as the theorem guarantees that incoherent credence functions will be) are irrational. As noted in Hájek (2008a), for the argument to work in favor of Probabilism, it better be the case that probabilistically coherent credence functions are not themselves so dominated. In fact, they are not—although this result is guaranteed by an appropriate choice of measure (the square distance, for instance), and it is not clear whether this isn't simply question-begging. Be that

[22] Strictly speaking, what can easily be seen is that they are closer measuring closeness by Euclidian distance. But the Brier score, which is what is usually used to measure distances in these contexts, preserves the order of Euclidian distance, and so the difference between them doesn't matter for the purposes of this demonstration.

as it may, the claim that credence functions that are accuracy-dominated are irrational is clearly false, as can be seen by reflecting on our case of T^*.

Remember that we are thinking of a specific probabilistically incoherent credence function, one that assigns credence less than 1 to a proposition that is in fact a tautology, but such that no ordinary human could tell that it is a tautology. How is it, exactly, that we can show that this credence function is accuracy-dominated? Suppose that our incoherent agent, perhaps having seen the above graph, recants and adopts coherent credences in P and $\neg P$, say $Cr_i(P) = .6$ and $Cr_i(\neg P) = .4$.[23] Still, when it comes to T^* our subject is still incoherent, assigning it credence less than 1: say, $Cr_i(T^*) = .7$. Let Cr_c be again a coherent function which agrees with Cr_i on P and $\neg P$, but which of course assigns 1 to T^*. How is it, exactly, that Cr_c dominates Cr_i? Well, T^* is true in both w_1 and w_2, and so Cr_i incurs an inaccuracy hit of 0.18 over and above the inaccuracies that are shared with Cr_c.

But notice what we did to figure out that result: we used our knowledge that T^* is a tautology, and thus that it is true in both w_1 and w_2. But, of course, someone whose credence function is Cr_i doesn't know that T^* is a tautology—that is the whole point of the example. That calculation, therefore, is not one she is in a position to carry out.

Why should it matter whether the subject whose credences we are considering is capable of carrying out the expected inaccuracy calculation or not? Remember why we moved from actual inaccuracy to accuracy-domination: we don't know which one is the actual world, and so the actual inaccuracy of a credence function has no implications for rationality. Whether a credence function is accuracy-dominated, on the other hand, is something that we can supposedly know a priori, for it involves simply a consideration of *all* the (relevant) possible worlds. But, in addition to not knowing which world is actual, we also don't know a priori exactly which propositions are true in which possible worlds. It's true that we can define possible worlds in terms of which *atomic propositions* are true in them, and so can (in principle) calculate the local inaccuracy of a credence function with respect to atomic propositions. But a subject unaware of the fact that T^* is a tautology will not, of course, know in which worlds T^* is true. For all he knows, it is indeed a tautology, true in every possible world, but it may also be a contradiction, true in no possible world, or a contingency, true in some but not all possible worlds. Therefore, the same reason we had to move from inaccuracy to accuracy-domination in the first place (namely, lack of knowledge of which propositions are true) is also a reason to disregard accuracy domination as giving us any meaningful information regarding the rationality of the credence in question.

[23] It's worth noting that it is not part of the Accuracy argument that subjects should adopt the credence distribution that is shown by the theorem to dominate the incoherent distribution in question.

2.7.2.3 Against Weak Probabilism

The case of T^* is one where the subject is unaware that a proposition that is in fact a tautology is a tautology, and relies on our computational limitations. For all I have argued so far, it is irrational to have a credence function you *know* to be probabilistically incoherent. But notice that this is not what Probabilism says: Probabilism says simply that rational credences should be probabilities, period. So, if this is the best that Dutch Book and accuracy-based arguments can provide, then they do not provide a vindication of Probabilism. In addition, I do not think that even the weaker claim, that a credence function you know to be incoherent cannot be rational, is true. Call this thesis "Weak Probabilism."

Consider my situation: I believe that a person with no hairs in his head is bald, that a hair cannot make the difference between being bald and not being bald, and that a person with around 100,000 hairs in his head is not bald. My credence function assigns a highish credence to each of those propositions. And yet, those propositions are an inconsistent triad—the truth of any two of them entails the falsity of the third one—and I know that they are an inconsistent triad. I also know that a probabilistically coherent credence function cannot assign a high credence to each proposition in an inconsistent triad (we'll see in a moment just how high those assignments can be).

Now, I know that many philosophers claim to have a solution to the sorites paradox. Most of them (though not all) involve denying, in some way or another, that a single hair cannot make the difference between being bald and not being bald. I don't know about you, but my credence in the correctness of any of these philosophical theories is approximately 0. If I were to indulge in cheap a priori psychology of philosophy, I would wager that not even the proponents of these theories have a high credence in them. But, be that as it may, I, at least, do not. Am I irrational for not having a low credence in the proposition that one hair cannot make a difference? Only philosophical hubris could lead someone to say so. It is a Moorean fact that no philosophical theory of vagueness is as remotely plausible as the claim that a single hair cannot make the difference between baldness and non-baldness. And, in this case, I know that my credence function is incoherent. And therefore I know, of course, that my credence function is accuracy-dominated, and this knowledge does not make it irrational.

But suppose that I yield to peer pressure and come to assign only a ⅔ credence to the proposition that a single hair cannot make the difference between being bald and not being bald. Even so, a probabilistically coherent credence function will allow me to assign a credence no higher than ⅔ to each of the other propositions! Consider the following stochastic truth-table, where P stands for the proposition that a person with no hairs in his head is bald, Q for the proposition that a person with 100,000 hairs in his head is not bald, and R for the proposition that a single hair cannot make the difference between being bald and not being bald:

P	Q	R	Cr
T	T	T	0
T	T	F	1/3
T	F	T	1/3
T	F	F	0
F	T	T	1/3
F	T	F	0
F	F	T	0
F	F	F	0

We are required to assign credence 0 to the row where all three propositions are true because they are an inconsistent triad. To maximize the credence in each proposition, we also assign 0 to the row where they are all false and where just one of them is true, effectively treating the case as if it were a fair three-ticket lottery, and then distribute our positive credence equally over the remaining three rows. The result is that each proposition gets assigned credence ⅔—and this is the maximum credence they can receive if we want them all to have the same credence. But whereas I may be pusillanimous enough to lower my credence in the proposition that a hair doesn't make a difference all the way down to ⅔ (although, to repeat, I do not think that I am rationally required to do that), I definitely have a higher credence in the propositions that Yul Brynner was bald and Brian May is not, and it would be preposterous to insist that I am thereby irrational.[24]

I don't think, therefore, that either traditional Dutch Book arguments or currently popular Accuracy-based arguments can do much to counter the *prima facie* implausibility of Probabilism. I conclude that Probabilism is false. That doesn't mean, however, that it is not fruitful to adopt it as a working hypothesis. In the next section I consider different ways of doing so, and give a lukewarm nod to one of them.

2.7.2.4 Modeling, Idealization, and Probabilism as Simplifying Assumption

So far, I have considered two arguments for Probabilism, the Dutch Book argument and the Accuracy-based argument, and I have argued that they both fail to

[24] Another way of making the same argument is to note that if your credences in P and Q are $1 - x$, then your credence in R is limited to $2x$. Thanks to an anonymous reader for Oxford University Press for this way of putting it.

show why having a credence of less than 1 in T^* is irrational. In this section I discuss some more considerations in favor of Probabilism. These considerations are not arguments for Probabilism, but rather defensive maneuvers designed to show that Probabilism's commitment to logical omniscience is not as bad as we might have otherwise thought.

I have been talking so far in terms of propositions and sets, but one could also define a probability function over sentences of an explicitly defined language. There are, of course, important differences between sentences and propositions. Whether a sentence expresses a tautology, for instance, is language-dependent, whereas whether a proposition is a tautology is not a language-dependent matter. Suppose now that we think of doxastic attitudes, including credences, as taking sentences as objects (or, perhaps, proposition-sentence pairs). Suppose, for instance, that we want to model a situation where we are rationally not certain of the logical relation between the sentence that it is not the case that neither Joe nor Mary will be at the party and the sentence that either Joe or Mary will be at the party. If the language we choose has a sentence representing that Joe will be at the party (J) and another representing that Mary will be at the party (M), then, given that $\neg(\neg J \wedge \neg M)$ is logically equivalent to $J \vee M$, Probabilism entails that any rational subject will assign them equal credence. But suppose that we choose a language with two different atomic sentences: J now stands for the proposition that it is not the case that neither Joe nor Mary will be at the party, and M for the proposition that either Joe or Mary will be at the party. Given that, taken as atomic sentences, J and M are logically independent of each other, Probabilism doesn't constrain the credences that a rational subject may assign to them. In particular, there is a probabilistically coherent probability function Pr such that $Pr(J) > Pr(M)$. Now, of course, the first kind of language is more perspicuous than the second in that, we think, it better reflects the metaphysics of the situation. But the idea is, precisely, that rational lack of logical omniscience depends on adopting a language that does not perspicuously reflect the necessity of all the propositions represented.[25]

That kind of response to the problem of logical omniscience, however, is necessarily limited. For while a judicious choice of language might obfuscate some of the tautologies we are rationally not certain of, any language, even the most impoverished one, will give rise to tautologies that we could not possibly be rationally required to be certain of. For consider the fact that the language in question will have to be closed with respect to conjunction, negation, and disjunction. This means that, even for a language with a single atomic sentence P, we can embed P in a mess of $\neg$s, $\wedge$s, and $\vee$s so complicated that no normal human can be expected to tell whether the resulting sentence is a tautology or not.

[25] For this kind of approach to the problem of logical omniscience see Garber (1983).

But there is another way in which the difference between sentences and propositions may help with the problem of logical omniscience. If we focus on a set-theoretic probability framework, then the bearers of credence are not linguistic entities like sentences but sets of events or possibilities (propositions, at least on some accounts). When we ask whether you assign credence 1 to all the tautologies, then, the answer will be "Yes," because there is only one tautology, and you are of course certain of it. Now, maybe we can access possibilities only through a linguistic veil, so to speak, which means that, strictly speaking, we may assign one credence to a possibility under one linguistic guise and a different credence to the same possibility under a different linguistic guise. If *this* is what the problem of logical omniscience amounts to, though, it is not clear that it is a problem for Probabilism set-theoretically defined, for Probabilism so defined is silent on the linguistic guises of possibilities. Of course, this may just mean that a set-theoretic approach to credences is wrong, because credences take as objects things with more structure than sets of possibilities, whether we want to identify those things with sentences or not.

A different approach to the problem of logical omniscience altogether is to say that Probabilism is required only for ideally rational agents, and that although all of us may fail to be logically omniscient, this is not true of ideally rational agents. This approach, however, is unsatisfactory insofar as it is not clear why logical ignorance is any different from, say, chemical ignorance. It is obviously not a requirement of rationality that one should know the chemical composition of water, so why would it be a requirement of rationality that one should recognize a recondite tautology as such? Gilbert Harman has in effect argued in this way against the idealization move (see, for instance, Harman (1986)). The rhetorical question is the following: if omniscience in general is not a requirement of even ideally rational agents, why would logical omniscience be any different? Christensen (2004) has responded that the difference is that there is a clear distinction between knowing facts and knowing how to reason, and that logical omniscience, while an idealization, corresponds to the "knowing how to reason" side of the distinction. However, Harman's point was that knowing that, through a complex reasoning, a certain proposition Q follows from another P, and using that knowledge to infer Q from P is not different from knowing that water is H_2O, and using that knowledge to infer that the glass contains H_2O from the proposition that the glass contains water. It is true that one of these facts is, in a sense which is of course hard to specify, "more contingent" than the other, but why would that make a difference? Perhaps the idea is that logical facts are knowable a priori—but it is hard to see why this is relevant when some of them are clearly not knowable (a priori or otherwise) *by us*. Moreover, as Christensen (2007a) argued, even ideally rational subjects have reason to believe that they are not rational, which arguably means that, even if they do in fact assign probability 1 to all the tautologies, they rationally shouldn't.

In a similar context, Robert Stalnaker (1991) distinguished four reasons to idealize: first, we may idealize because we want to get at underlying mechanisms; second, because we want to simplify; third, because, as Christensen (2004) does, we may take the idealization as a normative ideal; and, finally, we may want to idealize because it's the best we can do. Let's suppose that we take Probabilism to be not the literal truth but an idealization. For which of these four reasons would we want to idealize in the direction of Probabilism? I have already argued that Probabilism is not a normative ideal, so the third reason does not apply in this case. What about the other three?

In assuming Probabilism, do we idealize because we want to get at underlying mechanisms? We would if we thought that, for instance, the only reason why ordinary subjects do not assign maximal credence to every tautology is because of computational limitations. This does undoubtedly happen many times. But, as I argued in section 2.7.2.3, we may *knowingly* and rationally fail to be probabilistically coherent. In that case, our probabilistic incoherence is not due to the interference of our computational limitations on an underlying mechanism that would yield probabilistic coherence. Rather, the underlying mechanism itself correctly yields a probabilistically incoherent credence distribution. Therefore, we should not idealize for that first reason.

The remaining two reasons are related: we may want to idealize to simplify the treatment of the phenomena under study, and one excellent reason to want to simplify that treatment is that the underlying phenomena are so complicated that nothing but a simplified treatment will produce understanding. One obvious simplification in the idealization represented by Probabilism is that we simply ignore the acquisition of logical knowledge.[26] But, as Stalnaker himself notes, the acquisition of logical knowledge cannot be marginalized as only a logician's business. Every time we *reason* deductively we acquire logical knowledge, even when the subject matter of the reasoning has nothing to do with logic. When we take Probabilism as an idealization we are not, therefore, merely setting aside a small area of epistemology with limited application, but the whole faculty of reasoning, which has universal application.

We can take the idealization in Probabilism, then, as a model which is not concerned with deductive reasoning, but rather with what the result of such reasoning should look like (both before and after acquiring some evidence).[27] But why should we simplify in this direction, ignoring reasoning altogether? As I anticipated, one possible answer to this question is that reasoning might prove intractable to standard modelling techniques. There is certainly not *one* way in which we reason, and it might well turn out to be the case that there are no

[26] The attempt from Garber (1983) to model the acquisition of logical knowledge within a Bayesian framework is, as mentioned before, necessarily limited.

[27] It is in this connection useful to remember that Conditionalization itself (examined in the context of a normative interpretation in the next section) is not a model even of inductive reasoning, but only a description of what your credences should look like after such reasoning is done.

unifying principles governing the (infinitely?) many ways in which we reason. If this kind of particularism turns out to be true, then attempting to model them faithfully might resemble a Borgesian nightmare.[28]

But perhaps all of that is unduly pessimistic, and there will turn out to be unifying principles of reasoning amenable to theorization. If so, then so much the better. We can still consider Probabilism as a model which abstracts away from the details of such theorizing, at the same time that it imposes constraints on its results. Why would we want to do that? Because it is useful, for some purposes, to see which constraints are in place even when we set aside the intricacies of reasoning to concentrate on what the result of such reasoning should look like. Suppose, for instance, that we want to figure out the peculiar contribution that ignorance about non-logical facts makes to rational action—we want to figure out, for example, how your credence in the proposition that it will rain tomorrow combines with your preferences to issue in rational action. It would only confuse things if, in addition to your ignorance about the weather, we incorporate into the model your ignorance about logic. True, different actions might turn out to be rational if we assume that you are ignorant of some relevant logical fact, but this will not tell us anything about how the rationality of action varies with your uncertainty about the weather. The argument in the previous section that even credence functions that are known to be incoherent can be rational would apply even to agents for which reasoning is idealized away as suggested here. However, a similar point still applies: it would only confuse matters if we compose the problem of how to figure out what to do with the problem of how to solve the sorites or other philosophical paradoxes (which is not to say that those paradoxes are not interesting in their own terms, just that things would go better if we did not try to solve every problem at the same time).[29]

2.7.3 Conditionalization

Probabilism is a synchronic constraint on credences: it holds that, at any given time, the credences of a rational agent obey the axioms of probability. Many philosophers

[28] See Borges, "Of Rigor in Science" (my translation): "…In that Empire, the Cartographic Art reached such Perfection that the map of a single Province occupied a whole City, and the map of the Empire, a whole Province. With time, these Disproportionate Maps did not satisfy, and the Colleges of Cartographers raised a Map of the Empire which had the size of the Empire and which duly coincided with it.

Less Addicted to the study of Cartography, Subsequent Generations understood that the extensive Map was useless and not without Impiety they delivered it to the Inclemencies of the Sun and the Winters. In the deserts of the West there endure shattered Ruins of the Map, inhabited by Animals and Beggars; in the whole Country there is no other relic of the Geographic Disciplines." (Suárez Miranda, *Viajes de Varones Prudentes*, Libro Cuarto, Cap. XLV, Lérida, 1658)

[29] I return to the problem of logical omniscience in chapter 6, section 6.6.

working in the Bayesian tradition (that is to say, those philosophers who, broadly speaking, think that the axioms of probability have some substantive epistemological force) impose an additional constraint which relates the credences that an agent has at a time with the credences she has at a later time. Suppose that an agent's credences at a time t_1 are accurately described by the probability function P_{old}, and that this agent acquires some new evidence E at t_2, where $P_{old}(E) > 0$. According to Conditionalization, the agent's credence at $t2$ in any proposition P should then be given by a probability function P_{new} which bears the following relation to P_{old}:

Conditionalization: $P_{new}(P) = P_{old}(P|E)$

It is important to realize that the Conditionalization requirement is not a process requirement: it doesn't require the agent to *reason from* her credence in E to her new credence in P. Indeed, it is not even particularly enlightening to consider Conditionalization a diachronic constraint, for it doesn't say *anything* about how the agent should get from P_{old} to P_{new}. It is, rather, a comparative constraint, which merely requires as a necessary condition on rationality that a certain relation hold between P_{new} at t_2 and P_{old} at t_1.

The problems explored in section 2.5 for the definition of conditional probability as a ratio of unconditional probabilities arise anew in this context. Perhaps, it might be argued, the problems are not as pressing for the normative interpretation of probability as they are for the physical interpretation, for the normative interpretation does not apply to infinite sets of events. After all, the argument could go, we are finite beings, and so we can only consider, or at any rate assign credences to, finitely many events. It is not obvious this line of reasoning will work, however. For we can certainly consider the fact that the physical probability function is defined over some such set of events. And it has been argued that there are intimate connections between what I call physical and normative probabilities,[30] connections of a kind that would translate the issues with conditional physical probabilities to the case of normative probabilities.

When discussing Conditionalization for physical probabilities I noted that Conditionalization is *cumulative*, in the sense that, if a series of probability functions is such that any one of them results from conditionalizing the previous one on some proposition (except for the first one, of course, if there is one), then if a proposition is assigned probability 1 by any of the functions, then it is also assigned probability 1 by any of the succeeding functions. The same goes, of course, for probability 0. These extremes probabilities, it is sometimes said, are "sticky" if probability functions evolve by Conditionalization. This stickiness of extreme probabilities under Conditionalization was not a problem for physical probabilities, because once a fact obtains it cannot cease to obtain (although, of

[30] See the discussion of Lewis's "Principal Principle" in chapter 3.

course, a sentence that used to describe a fact may come to describe a non-fact). But for normative probabilities, the stickiness of extreme probabilities under Conditionalization is highly problematic. For, it seems, we may well forget some evidence that we once had, and, moreover, a proposition P can go from being part of our evidence to not being part of our evidence because of our acquiring some additional evidence Q which casts doubt on P. Conditionalization cannot deliver these results. I discuss these issues further in chapter 3, where I also discuss a variant of Conditionalization which does not suffer from this problem.

2.8 Traditional Decision Theory

So far, I have presented the mathematics of probability and considered the claim that our credences must obey it. That claim, I argued, is false, but assuming it might be harmless for certain purposes. The main purpose I have in mind is the development of decision theory, which presupposes Probabilism. In this section I present traditional decision theory—the theory of practical rationality as maximization of expected utility. We need to clarify what we mean by utility, its expectation, and the maximization thereof. But before that, we need to introduce the basic ontology of decision theory.

2.8.1 Actions, States, and Outcomes

On one way of presenting the traditional theory, we need to distinguish between the *acts* available to an agent at the moment of making the decision, the *states* the world might be in, and the *outcomes* that might result from performing one of the available acts while a certain state obtains. Thus, for example, it might be that there are two relevant acts available to you: to go to the park or to go to the movies. Two states might be relevant to your decision: it rains or it doesn't rain. Finally, there are the possible outcomes of your performing the available acts: you end up at the movies alone, you end up at the movies with your friend, you end up at the park alone, or you end up at the park with your friend. In table form:

	It rains	It doesn't rain
Go to the movies	Movies with your friend	Movies alone
Go to the park	Park alone	Park with your friend

The acts we need to consider, we said, are those that are "available" to the agent at the time of the decision. I am going to leave what it means for an act to be available at a pre-theoretic level. In particular, I am not going to get into such interesting issues as to whether acts that the agent cannot perform can ever be available in the relevant sense. These issues are in any case orthogonal to the ones investigated in this book.

Which states should we consider in the representation of a decision? This is a very important question, and detailed discussion of it will have to wait until chapter 3. For now, let us mention a pragmatic reason why some states need not be considered. If a state does not make a difference to the desirability of the acts in question for the subject, then it need not be considered. For instance, suppose that whether Magic Johnson eats a hotdog tomorrow makes no difference to the value you place in the different outcomes—that is to say, suppose that you are indifferent between being at the movies with your friend in a world where Magic Johnson ate a hotdog that day and being at the movies with your friend in a world where Magic Johnson did not eat a hotdog that day, and that the same goes for all the other possible outcomes. In that case, you need not consider how Magic Johnson's eating a hotdog would affect your decision—because it won't. This is not to say that, if a state does make a difference, then you should consider it. Sometimes you are justified in ruling out even states that make a difference—and this is the issue I take up in chapter 3.

Different authors treat acts, states, and outcomes differently. Savage (1954), for instance, treats states and outcomes as primitives, and defines acts as functions from states to outcomes. Jeffrey (1965), on the other hand, treats acts, states, and outcomes indifferently as propositions. Each choice brings with it some awkwardness—for instance, Savage's theory includes such weird acts as the one that assigns the outcome of a wet stroll to the state that it doesn't rain, whereas Jeffrey's theory has the consequence that we have preferences over how we shall act. These strange results need not detain us, however, for they are in large part technical requirements to prove the different "representation theorems" for the corresponding theories—on which more in section 2.10. If we abstract away from those requirements, we can think of our credences as distributed over the states, our preferences as defined over the outcomes, and the output of the theory as designating a set of acts as rational given those credences and preferences.

2.8.2 Utility

Rational action, according to traditional decision theory, is a function of your credences and your utilities. We know what your credences are, and we considered the norm that they should obey Probabilism. But what are your utilities? In traditional decision theory, a utility function represents your preferences over the set of outcomes. We will soon get more precise about this talk of "representation," but for now an intuitive understanding, according to which utilities stand for your preferences, will suffice. Your preferences can be represented with different kinds of *scales*. An *ordinal scale* is simply a ranking of outcomes, from the least preferred to the most preferred, with ties allowed. Thus, for instance, it may be that, in the case we were discussing above, the list of outcomes, in order from most to least preferred, is the following: park with your friend, movies with your friend, park alone, movies alone. With just this information, however, we will still not be in a position to determine which action is rational for you to perform. Going to the movies has something going for it, in that it will likely result in an outcome that is pretty good (is ranked second), but it also risks the worst possible outcome. Meanwhile, going to the park will likely result in a worse outcome than going to the movies, but it might also pay off and result in the best possible outcome of all. So, which is the rational action, going to the movies or going to the park? Armed with just an ordinal utility function to represent your preferences, we simply cannot answer that question.

What more do we need in order to answer that question? We need to know not only how you rank the possible outcomes, but also the relative distance between them. For instance, it may be that the difference between being with your friend at either the movies or the park is minimal, if being with your friend just overwhelms other qualities of the situation, whereas the difference between being at either place alone is huge. Or maybe it's the other way around, where if you are by yourself you barely care where you are, but being with your friend at the movies is so much better than at the park. Or maybe the difference between any two outcomes adjacent in the ranking is exactly the same. At any rate, this is the kind of information that we need in order to determine which action is rational. A utility function that represents not only your ranking of the outcomes but also the relative distance between them is known as a *cardinal* utility function.

We say that a utility function u *represents* your preferences over a set of outcomes if and only if, for every two outcomes A and B, $u(A) > u(B)$ if and only if you prefer A to B. A merely *ordinal* utility function gives relatively little information about your preferences. This is reflected in the fact that if a utility function represents your preferences in an ordinal scale, then any transformation of that utility function that preserves the order will represent your preferences as well as the original function. Thus, a function that assigns to the four outcomes the numbers 1, 2, 3, and 4 (in that order) will represent your preferences ordinally just as well as a function that assigns them -199, -1, 3, 3,000. A *cardinal* utility function, by contrast, gives more information. But even a cardinal utility function is not unique, in that if u is a cardinal utility function, then there will be transformations of u that preserve the differences between outcomes and, thus, represent the same information about the underlying preferences. These transformations are called "positive linear" transformations, and they have the following form: $u'(x) = au(x) + b$, where a is positive. We can see these transformations as changing the scale and the unit in which the underlying preferences are measured, much as Celsius and Fahrenheit are different scales for measuring the same underlying temperature.

2.8.3 Expectation

Mathematically, the expectation of a variable over a set is a weighted average of the values the variable takes over members of the set, where the weights come from some probability function. Let us unpack this idea, starting with an example.

You are going out for drinks. You consider how many glasses of wine you will have. You think it's somewhat unlikely you will have only one glass, as likely as not that you will have exactly two, just as likely that you will have three as that you will have one, and very unlikely that you will have four. You are sure you will have at least one glass, and no more than four. Let us say that, somehow, you manage to quantify the probabilities in question, as follows: you assign 0.2 to the outcome where you have exactly one glass, 0.5 to having two glasses, 0.2 to having three glasses, and 0.1 to having four. If you had to summarize in one number how many glasses of wine you expect to have, what would that number be? One possibility is to perform a simple average over all the numbers you think are possible, with the result being 2.5. But performing a simple average is, in effect, to assume that there is an equal chance of your drinking each number of glasses—when you think that is not case. Better, then, to perform a *weighted* average, where each number of glasses of wine is multiplied not by ¼, but by whatever you think the probability is that you will have that many. The result, in this case, is not much different (2.2), but the conceptual difference is nevertheless important.

What does this number, 2.2 in this case, represent? It is certainly not, despite the way I formulated the question, the number of glasses of wine you *expect* to have—let us suppose that you always finish drinking a glass once you start. If the expected value of a variable is not the value you expect it to take, what is it? The expected value is what the average of the value would approach if the experiment in question (in this case, going out for drinks) were to be repeated indefinitely (this the Law of Large Numbers).

2.8.4 Maximization of Expected Utility

Let us suppose that your preferences can be represented by a cardinal utility function, and that your credences obey Probabilism. Let us also suppose that the states relevant to your decision are a partition.[31] For concreteness's sake, let us suppose that the following are your utility and credence functions:

$u(\textit{movies with your friend}) = 7$
$u(\textit{movies alone}) = 1$

[31] That is to say, exactly one of the states is guaranteed to obtain. If the states that are relevant are pairwise incompatible but not jointly exhaustive, we can always recover a partition by adding a "none of the above" state.

$u(\textit{park with your friend}) = 15$
$u(\textit{park alone}) = 6$
$Cr(\textit{rain}) = .7$
$Cr(\textit{no rain}) = .3$

We can also represent the same information by adding it to our table:

	.7	.3
	It rains	It doesn't rain
Go to the movies	Movies with your friend **7**	Movies alone **1**
Go to the park	Park alone **6**	Park with your friend **15**

Notice that, given that it rains, you'd be better off going to the movies, whereas, given that it doesn't rain, you'd be better off going to the park. This means that neither of your available actions is *dominated* by the other. If one of the acts were dominated (if, say, $u(\textit{park alone}) = 7$, which means that going to the movies is *weakly dominated* by going to the park), then the decision would be easy: the only rational act is the non-dominated one. The justification for declaring dominated acts irrational is that, no matter which state of the world obtains, you would not be worse off performing a non-dominated act than performing a dominated one, and, for some state of the world, if it obtains then you would be better off performing a non-dominated act than a dominated one.[32]

But what is the rational action when more than one act is not dominated, as in our example? Traditional decision theory says that the rational acts are those that maximize expected utility. That is to say, the rational acts are those whose expected utility is not surpassed by any other act. In our example, the expected utility of going to the movies is $(7 \times .7) + (1 \times .3) = 5.2$, whereas the expected utility of going to the park is $(6 \times .7) + (15 \times .3) = 8.7$. The result is that the rational act is going to to the park. Generalizing:

$$EU(A) = \sum_i u(A \wedge S_i)Cr(S_i)$$

Notice that the act that traditional decision theory says is the rational one in this case is one which, you think, will most likely result in a suboptimal outcome: being

[32] That is the case of weak domination. If an act is strongly dominated, then performing a non-dominated act results in a better outcome no matter which state of the world obtains. The theorem at the heart of the accuracy-based argument examined before establishes that a probabilistically incoherent credence function is always dominated in the expectation of its accuracy by a coherent credence function.

at the park alone when you could have been at the movies with your friend, which you prefer. You think it's likely that it will rain, in which case going to the movies will have the best outcome. But what matters here are the *differences* between your preferences. Sure, you prefer being at the movies with your friend to being alone at the park, but your preference for being at the park with your friend over being alone at the movies is much stronger. The result is that, although you would be doing something which will (according to your own lights) likely result in a suboptimal outcome, this risk is more than outweighed by the probability (small, but significant enough) that you will get your most preferred result—being at the park with your friend. That is what a cardinal utility function allows us to do: to measure the difference between preferences. We need to ask, therefore, where that cardinal utility comes from.[33]

2.9 Where do Cardinal Utilities Come From?

We need, then, a cardinal utility function to calculate in a meaningful way the expected utility of the acts. But where does such utility function come from? We can, to be sure, form a ranking of all the possible outcomes, according to which we preferred to which ones. In fact, suppose we represent our preference relation by $\preccurlyeq$, where $A \preccurlyeq B$ means that B is not preferred to A—we can then define strict preference, $\prec$, as follows: $A \prec B$ *if and only if* $A \preccurlyeq B \wedge \neg(B \preccurlyeq A)$. Then, in order for us to be able to generate that ranking of outcomes, our preferences must satisfy the following axioms:

Completeness: For any A and B, either $A \preccurlyeq B$ or $B \preccurlyeq A$.

Transitivity: For any A, B and C, if $A \preccurlyeq B$ and $B \preccurlyeq C$. then $A \preccurlyeq C$.

But a preference relation that is only guaranteed to satisfy completeness and transitivity can meaningfully be represented only by an ordinal utility function—the only meaning the numbers assigned by the function have is the order they induce. We saw, however, that we need a cardinal utility function for the calculation of expected utility to make sense.

A famous theorem by von Neumann and Morgenstern (1944) lays down further axioms governing the preference relation which guarantee that the

[33] The version of expected utility maximization presented in this section is a Savage-style theory, where the states are assumed to be probabilistically independent of the acts. A Jeffrey-style theory lifts this assumption, calculating expected utility relative to the *conditional* probability of the state given the act, and not the unconditional probability of the state, as in Savage's theory. Jeffrey's theory, however, suffers from the Newcomb problem—see Nozick (1981). So-called "causal" decision theories (as opposed to Jeffrey-style "evidential" decision theories) replace the conditional probability by the probability of the subjunctive conditional linking the act with the state—see Gibbard and Harper (1978). For a criticism of causal decision theory see Egan (2007). I follow a Savage-style decision theory not because I think it is the right kind of decision theory, but rather because it is in some ways easier to follow, and the issues it suffers from are orthogonal to the ones I discuss in this book.

preferences will be representable by a cardinal utility function. The basic idea behind the theorem is that, in addition to simple outcomes like being at the movies with your friend, you must also have preferences over *lotteries* between these simple outcomes. Thus, let *(A, p, B)* be a lottery that results in outcome *A* with probability *p* and in outcome *B* with probability $1 - p$. Then, for the theorem to work, you should be able to order not only *A* relative to *B*, but also any lottery involving *A* and *B* relative to each other and to *A* and *B* themselves. More generally, let *S* be the set of "basic outcomes"—that is to say, those outcomes that are not themselves lotteries. Then we recursively define a set S' as follows: S' is the set that contains *S* and also contains, for every *A* and *B* in S' and any $p \in [0, 1]$, the lottery (A, p, B). Von Neumann and Morgenstern showed that if your preferences over S' satisfy certain axioms, then there is a cardinal utility function (unique up to positive linear transformations) that represents your preferences (in the sense previously defined) and that has the expected utility property. This last clause means that, for any lottery, the utility of that lottery is the expected utility of the lottery relative to the probability in question.

The axioms include the completeness and transitivity axioms just mentioned, this time applied to the set of all lotteries. They also include, however, further axioms which underlie the cardinality of the utility function. They are the following (where $A \sim B$ means that the agent is indifferent between *A* and *B*—i.e., both $A \preccurlyeq B$ and $B \preccurlyeq A$:)

Continuity: If $A \preccurlyeq B \preccurlyeq C$, then there is a $p \in [0, 1]$ such that $(\mathrm{A}, \mathrm{p}, \mathrm{C}) \sim \mathrm{B}$.

Independence: If $A \preccurlyeq B$, then for any C and any $p \in [0, 1]$: $(A, p, C) \preccurlyeq (B, p, C)$.

The continuity axiom means roughly that no outcome is so bad that you are not indifferent between the status quo and a lottery that gives some probability (no matter how small) to the bad outcome, but also some chance of an outcome better than the status quo. The independence axiom means that your preference over lotteries should not be affected by what they have in common, only by what they differ in.

If your preferences do satisfy all these axioms, then there is a utility function *u* which represents them and which has the expected utility property. The received opinion is that, given this theorem, preferences have psychological reality, and utilities are simply constructed from these preferences.

2.10 Representation Theorems

The von Neumann-Morgenstern theorem examined in the previous section is a special case of a more general phenomenon. There are other theorems which show that, if your preferences satisfy certain additional axioms, then you can be represented as having credences that satisfy a certain probability function and a certain utility function such that you maximize expected utility relative to those functions. That is to say, if your preferences satisfy the axioms, then there is a unique probability function *Cr* and a utility function *u* such that you prefer action *A* to action *B* if and only if the expected utility of performing action *A* (relative to *Cr* and *u*) is greater than that of performing action *B* (relative to *Cr* and *u*).

There are two important questions that we should ask about representation theorems. First, how could they work? In particular, given that the axioms refer exclusively to your preferences, and not to your credences, how can they give rise to a unique probability function? Second, given that the theorems do work, what is their philosophical import?

The first question can be answered by noticing that, although the axioms do indeed pertain to preferences, your opinions can be gleaned from some of your preferences. In particular, consider a special kind of lottery between outcomes, lotteries where the probability of the outcomes is not given explicitly but through the truth-value of a proposition. Consider, for instance, the lottery that gives you outcome *A* if it rains tomorrow, and outcome *B* otherwise (where you prefer *A* to *B*). If you prefer that lottery to the one that gives you outcome *A* if it doesn't rain tomorrow, and *B* otherwise, then you think it is more likely than not that it will rain tomorrow.

There are two interesting issues that representation theorems give rise to. The first one is that some philosophers have used representation theorems as arguments for Probabilism. The idea is that the axioms are rationality requirements on preference (and, indirectly, on doxastic attitudes), and given that anyone who satisfies the axioms has opinions that can be represented by a probability function, it follows that Probabilism is a requirement of rationality.

But the argument is invalid. Just because something can be represented as having a certain property it doesn't follow that it has that property. For the purposes of some calculations, for instance, planets can be represented as being point-sized—but, of course, planets are not point-sized. Analogously, just because for the purposes of the calculation of expected utility the opinions of an agent who satisfies the axioms can be represented by a probability function doesn't mean that the opinions themselves are probabilistically coherent.

Moreover, as Zynda (2000) has shown, the representations are not unique. Any subject who can be represented as maximizing expected utility relative to a probability function *Pr* and a utility function *u* can be represented as maximizing *schmexpected* utility relative to an alternative utility function u' and a probabilistically incoherent credence function Pr'. If mere representability were sufficient for existence, then subjects whose preferences satisfy the axioms would have both coherent *and* incoherent credence functions.

Although they do not give us a good argument for Probabilism, representation theorems do give rise to an interesting question. Usually, on the basis of the von

Neumann-Morgenstern representation theorem, preferences are taken to be basic and psychologically real, whereas utilities are taken to be derivative and merely constructed from the underlying preferences. Why don't we take probabilistically coherent credence functions to be similarly constructs that merely represent an underlying psychological reality? The underlying psychological reality, in the case of credences, would be a relation of comparative belief. Why not take these comparative beliefs to be basic, and the probability function merely one of many possible representations of it? This is Zynda's view in the article just cited.[34] This position bypasses some of the issues related to Probabilism, although not all of them—for instance, a comparative analogue of normality is still required, which raises the same issues discussed in relation to the probabilistic requirement of normality.

2.11 Conclusion

In section 2.7 I introduced normative probabilities as the claim that a necessary condition on the rationality of our credences is that they respect Probabilism. Then, in section 2.7.2, I criticized Probabilism as well as the Dutch Book argument and the accuracy-based argument for that view. I ended that section by endorsing the idea that we can take on Probabilism as a harmless simplification meant to isolate the impact of our empirical ignorance on the rationality of our actions. In 2.7.3 I introduced Conditionalization as another necessary condition on the rationality of our credences. Whereas Probabilism is a synchronic constraint which applies to our credences at any given time, Conditionalization is a comparative constraint which applies to credences held at different times.

I will take subjective Bayesianism to be the view that Probabilism and Conditionalization, together, are not only necessary but also sufficient conditions on the rationality of credences. That is to say, according to subjective Bayesianism, as long as your credences at any given time satisfy Probabilism, and as long as they evolve according to Conditionalization, they are rational. The next chapter is dedicated to examining the shortcomings of subjective Bayesianism and presenting objective Bayesianism as an alternative.

[34] But cf. Meacham and Weisberg (2011) and, in reply, Stephánsonn (2017).

3
Objective Bayesianism and Knowledge-First Epistemology

3.1 Introduction

In the last chapter I introduced the mathematics of traditional decision theory. That theory requires credences to be probabilistically coherent. I argued that this requirement is wrong (rational credences can be probabilistically incoherent, and even known to be so), but I also conceded that, in some contexts (for instance, in the context of examining the peculiar impact of empirical uncertainty on practical rationality), it may well be harmless to assume that rational subjects have probabilistically coherent credence functions. Let me say a few more words here about why I think that the assumption is indeed harmless as far as the issues discussed in this book go. The overall view presented in this book has it that false beliefs can play central roles in epistemology and decision theory. A main argument for that view moves from considerations regarding practical rationality to considerations regarding epistemology: I will argue that certain actions are rational, and that the rationality of those actions presupposes the rationality of some false beliefs. In making this argument, I assume that the credences in question in determining the rationality of action are probabilistically coherent. Lifting this assumption would not change the fundamentals of the argument—it would just make application of traditional decision theory impossible, and therefore judgments about which actions are rational messier.

Subjective Bayesianism, presented at the end of last chapter, assumes that the requirement of probabilistic coherence is not only necessary for the rationality of credences, but also sufficient. In this chapter I argue against Subjective Bayesianism, and present Objective Bayesianism as an alternative. Objective Bayesianism is not just the negation of the subjective variety. The negation of Subjective Bayesianism just has it that not any probabilistically coherent credence distribution is rational. What I here call "Objective Bayesianism" lies at the other extreme of the spectrum from Subjective Bayesianism: it is the claim that, given some evidence, exactly one credence function is rational.[1] Objective Bayesianism is seen

[1] More precisely, Objective Bayesianism has it that there is exactly one rational ur-prior (the notion of an ur-prior is clarified in the next section).

Being Rational and Being Right. Juan Comesaña, Oxford University Press (2020). © Juan Comesaña.
DOI: 10.1093/oso/9780198847717.001.0001

in many circles as a deeply implausible project, but I will defend it in this chapter. My defense of Objective Bayesianism, however, will be conditional. Given that I believe that rational credence functions need not be probabilistically coherent, I don't really believe in Objective Bayesianism. But I do believe that Objective Bayesianism is closer to the truth than the subjective variety. Given that we are assuming that probabilistic coherence is a harmless idealization for the purposes that we are interested in, more than probabilistic coherence is required of a rational credence function.

Chapter 4, therefore, will assume that Probabilism is a harmless idealization, but that some form of Objective Bayesianism is correct (given that first assumption). Lifting these two assumptions would only make my argument easier. For suppose that we lifted the assumption of Probabilism. That opens the door to credence functions like the one introduced in the last chapter, where high credence is assigned to each of a set of incompatible propositions. At least one of those propositions must be false. Of course, there is still a gap from here to the conclusion that there can be false rational beliefs, but we can at least glimpse the contours of an argument that is closed off once we assume Probabilism. And suppose that we lift the assumption of Objective Bayesianism, leaving us with the subjective variety instead. It is an immediate consequence of Subjective Bayesianism that it is rational to give credence 1 to falsehoods. Again, there is a gap from here to the conclusion that there can be false rational beliefs (although, in this case, it is a gap that I will close in the next chapter), but, also again, there is a line of argument here that is simply unavailable if we reject Subjective Bayesianism. Thus, although I think that Objective Bayesianism, interpreted as the literal truth, is wrong, I am only making my life harder by assuming it as background of the arguments in the following chapter.

This chapter unfolds as follows. I start by presenting the notion of an ur-prior, and then explain what the subjectivity of Subjective Bayesianism amounts to in terms of ur-priors. I then examine Lewis's Principal Principle as an initial move away from extreme versions of Subjective Bayesianism. After that I discuss in some detail Carnap's program of developing a form of Objective Bayesianism in purely syntactical terms. That project fails, as demonstrated by Goodman's "grue" problem. It has been widely assumed that the failure of Carnap's project portends the failure of Objective Bayesianism more generally. This idea has been recently developed in the form of a clear argument by Michael Titelbaum. I examine and reject Titelbaum's argument. Objective Bayesianism, freed from Carnap's strictures, is possible. In its extreme form for which I argue here (always under the assumption of Probabilism as a simplifying idealization), Objective Bayesianism is the claim that exactly one ur-prior is rational. But even after we move from Subjective to Objective Bayesianism two important concerns remain. The first one has to do with the "stickiness" of extreme (0 or 1) probability assignments under Conditionalization. I present Ur-prior Conditionalization as an updating rule which has the advantage of not making extreme probability assignments sticky. The second concern is that Bayesianism, of either variety, is simply silent regarding the central notion of evidence. Timothy Williamson has argued for an elegant answer to that concern: a subject's evidence at a time consists of all the propositions the subject knows at that time. I end this chapter by presenting this view in the context of Williamson's wider knowledge-first epistemology project. This sets the stage for the argument of next chapter, where I argue against this Williamsonian version of Objective Bayesianism.

3.2 Ur-Priors

It will be useful to introduce the notion of an ur-prior in order to understand both the objectionable subjectivity of Subjective Bayesianism as well as the alternative offered by Objective Bayesianism.

Suppose that a subject abides by both Probabilism and Conditionalization—i.e., the credences of this subject at any given time obey the probability calculus, and, whenever the subject receives some new evidence, he updates his credences by conditionalizing on this new evidence. One issue with Subjective Bayesianism is that it is silent on what the evidence is—more on this in section 3.10. But, for now, let us leave this at an abstract level. Let us then assume that $t_1 \ldots t_n$ are the different times at which the subject receives new evidence, that Cr_i is the subject's credence function at t_i, and that E_i is the evidence the subject receives at t_i. Each Cr_i, then, is a probability function, and for any $i \geq 1, Cr_i(-) = Cr_{i-1}(-|E_i)$.

Whenever those assumptions obtain—whenever a subject abides by the strictures of Subjective Bayesianism—there will be a credence function Cr_u such that each $Cr_i(-) = Cr_u(-|E_1 \wedge \ldots \wedge E_i)$. That is to say, there will be a credence function Cr_u such that every credence function the subject has at any time t_i is the result of conditionalizing Cr_u on the conjunction of all the evidence the subject has had up until t_i. For this reason, some authors call Cr_u an ur-prior.

Let me illustrate the existence of ur-priors with an example from Titelbaum (forthcoming). Suppose that I roll a dice outside of your view. Then, at t_1, I tell you that the dice came from a reputable casino. At t_2 I tell you that the roll came up even. Finally, at t_3 I tell you that the roll came up a prime number. We still do not have a theory of evidence to plug into the Bayesian framework, but I will assume that your evidence at t_1 is that the dice came from a reputable casino (C), at t_2 that the roll came up even (E), and at t_3 that it came up prime (P). Let us also suppose that your credences at each of these times are the ones represented in the following stochastic truth-table:

C	P	E	Cr_1	Cr_2	Cr_3
T	T	T	1/6	1/3	1
T	T	F	1/3	0	0
T	F	T	1/3	2/3	0
T	F	F	1/6	0	0
F	T	T	0	0	0
F	T	F	0	0	0
F	F	T	0	0	0
F	F	F	0	0	0

Which is the Cr_u whose existence is guaranteed and which bears the required relation to the different Cr_i? In fact, there is no such thing as the unique Cr_u—there will be infinitely many. To see why, consider first a somewhat natural Cr_u. Suppose that, before I told you that the dice came from a reputable casino, your credence that it did was 0.5. You also don't have any opinions as to how dice that do not come from reputable casinos are biased, and so your credences as to how a roll from a dice will come up are insensitive to whether the dice comes from a reputable casino or not. In that case, the Cr_u is as in the following table:

C	P	E	Cr_u	Cr_1	Cr_2	Cr_3
T	T	T	1/12	1/6	1/3	1
T	T	F	1/6	1/3	0	0
T	F	T	1/6	1/3	2/3	0
T	F	F	1/12	1/6	0	0
F	T	T	1/12	0	0	0
F	T	F	1/6	0	0	0
F	F	T	1/6	0	0	0
F	F	F	1/12	0	0	0

What I did in order to get that Cr_u was to divide all the non-zero Cr_1 credences in complete state-descriptions in half, and assign the other half to the other state-descriptions in the same proportions. But I could have, for instance, assigned that other half in different proportions—for instance, by giving the whole 0.5 to the state-description where all propositions are false. We can also divide the original credences not just in half, but in whatever proportion we want. For instance, we can assign just ⅔ to ¬*C*, and ⅓ to *C*, while keeping the same proportions as Cr_1 for the non-zero state-descriptions:

C	P	E	Cr_u	Cr_1	Cr_2	Cr_3
T	T	T	1/18	1/6	1/3	1
T	T	F	1/9	1/3	0	0
T	F	T	1/9	1/3	2/3	0
T	F	F	1/18	1/6	0	0
F	T	T	1/6	0	0	0
F	T	F	1/6	0	0	0
F	F	T	1/6	0	0	0
F	F	F	1/6	0	0	0

The person whose credence function is represented in that table thinks that it is ⅔ likely that the dice is from a reputable casino. Unlike the previous person, however, he also has opinions as to how dice that do not come from reputable casinos are biased: he thinks that they are more likely to come up either 1 or 2 than any other number.

The only trick to get a Cr_u with the desired property (i.e., that any Cr_i is the Conditionalization of that Cr_u on the conjunction of all the evidence had up until t_i) is to assign to the first four rows (the state-descriptions where *C* is true) whatever proportion *n* we want, to multiply the non-zero credences assigned by Cr_1 by *n* to get the credences assigned to those states by Cr_u, and then assign the remaining credence over the ¬*C* states as we wish. In particular, if we want we can make the Cr_u *regular*, i.e., we can make it assign positive numbers to every proposition.[2]

Ur-priors reflect the cumulative nature of Conditionalization, for, as I said, we can get any Cr_i by conditionalizing Cr_u on the conjunction of all the evidence the subject has accumulated up to t_i.[3] Ur-priors also reflect the commutative

[2] More on regularity in the next chapter.

[3] To say that evidence is cumulative is to say that once a proposition becomes part of a subject's evidence, it never ceases to be part of that subject's evidence. This is not a desirable property of evidence—see section 3.9.

nature of Conditionalization.[4] For notice that, regardless of the order in which we conditionalize Cr_u on E_1, E_2, and E_3, the result after conditionalizing on all three will always be Cr_3.

What do ur-priors represent? If they are regular, they need not represent the subject's credences at any given time, for there need not be any time in the subject's epistemic life where he didn't have some non-trivial evidence. Regular ur-priors may be seen as representing the *evidential principles* the subject abides by. It is perfectly possible for two different subjective Bayesian subjects to receive the same evidence and yet differ on which credences they adopt in light of that evidence. The immediate explanation of this is, of course, that the credence function they conditionalize the evidence on need not be the same for both of them. The difference in this prior credence function may, in turn, be explained by a difference in the evidence they received before the common piece of evidence. But it is also possible for two subjective Bayesian subjects to be evidential twins, i.e., to have received the same evidence throughout their lives, and still have different credence functions. When this happens, the ur-priors for the subjects will be different. These different ur-priors represent the different evidential principles which the subjects respect. Thus, maybe one of them gets more and more convinced that the next observed emerald will be green the more green emeralds he sees, and the other one gets more and more convinced that the next emerald will not be green given the same evidence.

I showed before that a sequence of credence functions related to each other by Conditionalization need not generate a unique ur-prior for the subject whose credences they are. This is because any given sequence will represent only a subset of every possible sequence of bodies of evidence (no sequence can represent every possible sequence of bodies of evidence because some of them are incompatible with each other—for instance, there is a sequence that starts with E and another one that starts with $\neg E$). What any given sequence determines is a subset of all the conditional probabilities in the ur-prior—in our example, it determines $Cr_u(-|C), Cr_u(-|E)$, and $Cr_u(-|P)$, but it doesn't determine, for instance, $Cr_u(-|C \wedge P \wedge \neg E)$. If we *are* given all the conditional probabilities, then a unique Cr_u is indeed determined. This is trivially the case, for all the conditional probabilities include the probabilities of every proposition conditional on a tautology, which are equivalent to the unconditional probabilities of every proposition. But there is also a non-trivial subset of conditional probabilities which determine the whole probability distribution. I illustrate first the case of a distribution over two atomic propositions, and then I generalize the result.

[4] To say that evidence is commutative is to say that the result of first receiving evidence E_1 and then evidence E_2 is the same as that of first receiving evidence E_2 and then E_1. Commutativity is usually seen as a desirable property of evidence—but see Lange (2000).

Consider first, then, a stochastic truth-table for two propositions, E and H:

E	H	Cr_u
T	T	a
T	F	b
F	T	c
F	F	d

Because we are assuming that Cr_u is a probability function, we know that $a+b+c+d=1$. Suppose now that we are given the following three conditional probabilities: $Cr_u(E \wedge H|E), Cr_u(E \wedge H|H), Cr_u(E \wedge H|E \leftrightarrow H)$. These conditional probabilities are the ratios $\frac{a}{a+b}, \frac{a}{a+c}, and \frac{a}{a+d}$. Thus, we have four unknowns (a, b, c and d) and four non-equivalent, non-contradictory linear equations. This system of linear equations therefore has a unique solution, and so we know the values of a, b, c and d. And because the credence for any proposition is the sum of the credences of each state-description where the proposition is true, we therefore have the unconditional credences for every proposition.

Generalizing, suppose that the credence function is defined over a Boolean algebra with n atomic propositions. This algebra determines a stochastic truth-table with 2^n state-descriptions (the S_is). Because Cr_u is a credence function, the sum of its values over all the state-descriptions is 1. Moreover, suppose that we are given values for the x_i in all the equations of the form $Cr(S_1|S_1 \vee S_i) = x_i$, for $2 \geq i \geq 2^n$. Together, these equations form a system of non-equivalent, non-contradictory 2^n linear equations with 2^n unknowns, and therefore the system has a unique solution. We can therefore compute the value for each state-description, and from them the unconditional credence in every proposition.

This result is surprising. In the Subjective Bayesian framework, conditional probabilities encode, as I said, the epistemic principles that the subject adopts—and the conditional probabilities determined by the ur-prior are the fundamental epistemic principles that the subject adopts. But it is also natural to think that epistemic principles determine which credence we should have in particular propositions only in conjunction with some evidence. An evidential principle by itself tells us which credences to adopt in reaction to some evidence, but not which credences to adopt in the absence of evidence—or so the natural thought goes. The determination of unconditional probabilities by conditional probabilities noticed in the previous paragraph, however, shows that this natural thought is wrong. It is, of course, a staple of Bayesian epistemology that *some* unconditional probabilities are constrained by conditional probabilities. For instance, a Bayesian solution to the grue problem (on which more in section 3.7) requires the agent to assign a higher unconditional probability to the hypothesis that all emeralds are green than to the hypothesis that all emeralds are grue, even in the absence of any evidence. But the point generalizes: the whole set of fundamental epistemic principles determines which credences we should give to *every* proposition in the absence of any evidence for it. It is impossible for two subjects to agree on the evidential impact of every bit of evidence but disagree on which credences to assign in the absence of evidence. The unique ur-prior which is determined by the whole set of conditional probabilities, then, encodes two things: the fundamental epistemic principles that the subject adopts, of course, but also the credences the subject would adopt towards any proposition a priori.

3.3 The Subjectivity of Subjective Bayesianism

The whole set of conditional probabilities, then, uniquely determines an ur-prior—and, of course, any ur-prior determines a complete set of conditional probabilities. According to Subjective Bayesianism, moreover, *any* probabilistically coherent ur-prior is rational. That means that Subjective Bayesianism is compatible with *any* way of accommodating evidence so long as it abides by Probabilism and Conditionalization, even ones that we would naturally consider irrational. This fact is so important that I will call it:

The Fundamental Theorem of Bayesian Epistemology: for any logically independent propositions E and H, and any real number n between 0 and 1, there are infinitely many probability functions Cr such that $Cr(H|E) = n$.[5]

Consider, for instance, the following case. There are a bunch of emeralds in a bag—they are all green. A subject with no prior knowledge of emeralds and their color will extract the emeralds one by one from the bag and note their color. Let E_i be the proposition that the emerald extracted from the bag at t_i is green. Let us consider the portion of a credence function which is defined over the E_i. I will assume (pending a theory of evidence on the part of the Bayesian) that the subject's evidence at t_n is the conjunction $E_1 \wedge \ldots \wedge E_n$. A natural credence function for such a subject to have is one where $Cr_u(E_{n+1}|E_1 \wedge \ldots \wedge E_n) > Cr_u(E_{n+1})$—that is, the subject's credence that the emerald to be extracted at t_{n+1} is green is greater conditional on all the previous emeralds being green than unconditionally. Maybe it is also natural for the rate of this increase in confidence to grow as n grows—that is to say, the more green emeralds the subject observes, the more convinced he becomes that the next one will also be green. But there are coherent

[5] Proof: Set the ratio of $E \wedge H$ to E to n (there will be infinitely many ways of doing this). This will "use up" some amount m of credence. Distribute the remaining $1 - m$ credence as you see fit among the remaining state-descriptions (and there will be, again, infinitely many ways of doing this).

credence functions where the inequality just mentioned is reversed—the subject becomes more and more convinced that the next emerald will *not* be green the more green emeralds he observes. Given that these are coherent credence functions, they are sanctioned as rational by Subjective Bayesianism, and thus someone whose credences are correctly represented by one of these is sanctioned as having rational credences by Subjective Bayesianism.

This is, of course, not news to proponents of Subjective Bayesianism. They see the Fundamental Theorem of Bayesian Epistemology as a feature. For many of us, however, it is a bug. The main argument in support of the Subjective Bayesian position on this issue has always been that the alternative—that there are further constraints on ur-priors besides the probabilistic coherence constraint—is indefensible. To be sure, some Bayesians are happy to incorporate further constraints into ur-priors, but the thought remains that to posit anything close to a unique rational ur-prior is a pipe dream. In the next section I briefly outline some of the constraints on ur-priors that philosophers have posited, and then I present Objective Bayesianism and tackle head-on what I take to be the main objection against it.

3.4 Further Constraints on Ur-Priors

In the previous chapter I introduced the probability calculus and briefly presented two interpretations of it: Physical and Normative probabilities. Physical probabilities, recall, are probabilities assigned to events by complete and correct basic scientific theories—our incomplete and probably incorrect (at least in the details) current scientific theories are defeasible guides to Physical probabilities. Normative probabilities, on the other hand, are rational credences—which Bayesians, at least, take to be governed by the norm that they be probabilistically coherent. Are there any connections between Physical and Normative probabilities?

David Lewis (1986) thought so. He put forward the *Principal Principle* about the relation between Physical and Normative probabilities. Remember that we are thinking of the Physical probabilities as determined by the ur-Physical prior (which encodes all the true natural laws).[6] Following Lewis, let's call this probability the "chance" of an event A, and let's represent it as follows: $Ch(A)$. The Principal Principle can then be formulated as a constraint on admissible normative ur-priors, as follows:

Principal Principle: $Cr_u(A|Ch(A) = x \wedge E) = x$, provided that E is admissible with respect to $Ch(A)$.

[6] I ignore in what follows the fact that these probabilities evolve by Conditionalization on the facts.

That is to say, the Principal Principle sets a constraint on normative ur-Priors according to which the conditional probability of any event given a particular kind of evidence about that event needs to be equal to the chance of that event. The evidence in question is a conjunction whose first conjunct is, precisely, the chance of that event, and whose second conjunct E is any additional evidence regarding the event in question, provided that this additional evidence is "admissible" with respect to the chance of the event. Obviously, whether the Principal Principle is plausible depends heavily on what makes for the admissibility of the additional evidence, and we will turn to that issue momentarily. But, forgetting for a moment about that additional evidence E (or restricting ourselves to the cases where it is null), what the Principal Principle says (given Conditionalization as an updating rule) is that if all the subject's evidence regarding A is that its chance is x, then the subject's credence in A should also be x. Thus, for instance, if your only evidence about a certain coin flip is that the chance of its having come up heads is $\frac{1}{2}$, then your credence that the flip came up heads should also be $\frac{1}{2}$.

Of course, we do not, in general, have evidence which settles the exact chances of events. But that doesn't make the Principal Principle inapplicable. We may have evidence about the possible chances of an event: we may know, say, that the chances of heads are either $\frac{1}{2}$ or $\frac{1}{3}$, and, moreover, we may come to rationally assign credence $\frac{2}{3}$ to the first possibility and $\frac{1}{3}$ to the second—that is to say, we may come to have evidence E such that $Cr_u(ch(A) = \frac{1}{2}|E) = \frac{2}{3}$ and $Cr_u(ch(A) = \frac{1}{3}|E) = \frac{1}{3}$. If so, the Principal Principle together with an application of the (conditional form of) the Law of Total Probability[7] entails that $Cr_u(A|E) = \frac{4}{9}$. That is to say, if all our evidence about A is E, then the Principal Principle entails that we should assign credence $\frac{4}{9}$ to A.

Let us now go back to the issue of admissibility. What Lewis has in mind here is the following kind of case: suppose that your evidence E about event A is that its chance is $\frac{1}{2}$ *and* that it obtained. That evidence entails A, and so, for any coherent Cr_u, $Cr_u(A|E) = 1$. Thus, it better not be the case that this evidence counts as admissible relative to $Ch(A)$, for then the Principal Principle will generate inconsistent recommendations. In general, admissible evidence will affect your credence in an event only *via* affecting its chances. Thus, any evidence which affects credences *directly* is, in the relevant sense, inadmissible. Notice that there is nothing wrong with inadmissible evidence—it just renders the Principal Principle inapplicable.

This is relevant to the issue of how much of a departure from Subjective Bayesianism the Principal Principle represents. Without any further constraints on what makes for the inadmissibility of evidence, the Subjective Bayesian is free

[7] The conditional form of the Law of Total Probability is the following:

$$Pr(A|B) = \sum_n Pr(A|B \wedge C_n)Pr(C_n|B)$$

to hold on to the Principal Principle without compromising one iota of his extreme subjectivism. For, without any further constraints, it is perfectly open to the Subjective Bayesian to adopt the following trivializing account of admissibility: evidence E is inadmissible with respect to $Ch(A)$ if and only if $Cr_u(A|ch(A) = x \wedge E) \neq x$.

In addition to the Principal Principle, further constraints on ur-priors have been advanced. van Fraassen (1984), for instance, advanced a *reflection principle*, according to which we should defer to our future credences in much the same way that the Principal Principle has it that we should defer to chances. Much of the discussion regarding the epistemic significance of disagreement can be formulated in terms of whether we should add further constraints on ur-priors regarding deference to epistemic peers.

There is a spectrum of Bayesian views, with Subjective Bayesianism at one end. Views which accept (non-trivial) versions of further constraints to the ur-priors move away from the Subjective Bayesian extreme. But even accepting all the principles discussed in this section will fall short of the other extreme, Objective Bayesianism, according to which there is only one rational ur-prior. A step in the direction of Objective Bayesianism is provided by the Principle of Indifference, which I discuss next.

3.5 The Principle of Indifference

The venerable Principle of Indifference, which dates back to Laplace, promises to deliver constraints on ur-priors so strong as to determine a unique rational ur-prior. Remember that the ur-prior encodes two different things: a set of evidential principles (in its conditional probabilities) and the intrinsic credibility of any proposition in the absence of any evidence. The Principle of Indifference is intended precisely to determine the rational credence in a proposition in the absence of evidence. Thus, the Principle of Indifference, if successful, will completely determine a unique ur-prior as rational.

In its usual formulation, the Principle of Indifference says that, in the absence of evidence, credences should be distributed uniformly over the members of a partition. Applied to ur-priors, the Principle is the following:

Principle of Indifference: If $P_1 \ldots P_n$ form a partition, then, for all i and j between 1 and n, $Cr_u(P_i) = Cr_u(P_j)$.

The propositions that I have no sisters, that I have exactly one sister, and that I have more than one sister form a partition. Thus, an application of the Principle of Indifference has it that a rational ur-prior would assign credence $\frac{1}{3}$ to each such proposition. This example highlights an obvious problem with the Principle of

Indifference. For here is another partition: I have no sisters, I have exactly one sister, I have exactly two sisters, and I have more than two sisters. An application of the Principle of Indifference to *this* partition yields the result that a rational ur-prior would assign credence $\frac{1}{4}$ to each such proposition. But then, what credence should a rational ur-prior assign to, say, the proposition that I have no sisters, $\frac{1}{3}$ or $\frac{1}{4}$—or one of the infinitely many other credences delivered by applying the Principle of Indifference to one of the infinitely many other partitions? Unconstrained, the Principle of Indifference is simply contradictory.[8]

Now, maybe there are some preferred partitions. For instance, suppose that your credences are defined over a set whose atomic propositions are $P_1, \ldots, P_n$. This generates 2^n state-descriptions. One natural partition over which to apply the Principle of Indifference would be precisely the one given by the different state-descriptions. Each of those state-descriptions is a specification of a way the world could be which is complete as far as the descriptive power of the propositions in question goes. This is precisely one of the guiding ideas in Carnap's program of devising a logic of confirmation, so I turn to it now.

3.6 Carnap on Confirmation

When does a body of evidence count in favor of a scientific hypothesis? Mid-twentieth-century philosophers of science (like Hempel and Carnap) addressed this question within a formal setting. Their goal was to discover (or create) an "inductive logic"—that is to say, a theory that answered that question with the same degree of precision that traditional (deductive) logic answered the question of what it means for a set of sentences to entail another. Although their focus was on the confirmation of scientific hypotheses by scientific evidence, their research applies equally well to the justification of any kind of proposition by any body of evidence. We are particularly interested in Carnap's development of the project, for he explicitly formulated it within a probabilistic framework.[9]

We can begin with a relative notion of confirmation, confirmation with respect to a probability function Pr. Carnap's project will then be that of moving from this relative notion of confirmation to an absolute one. But even within this relative notion, Carnap makes an important distinction between two kinds of confirmation: confirmation as *firmness* and confirmation as *increase in firmness*. E confirms H in the firmness sense if and only if $Pr(H|E) > r$, where r is some threshold larger than 0.5. Thus, confirmation as firmness compares the conditional

[8] Similar, but perhaps harder, problems arise for credences defined over an infinite space. Bertrand's Paradox asks how likely it is that a randomly chosen chord of a circle will be larger than the side of an equilateral triangle inscribed on the circle. Depending on how the set of chords is partitioned (for instance, by the chord's endpoints, by their midpoints, etc.) the answers are different.

[9] In this section I rely heavily on Fitelson (2005), which provides a nice overview of Carnap's project.

probability of H given E with a constant. On the other hand, E confirms H in the increase in firmness sense if and only if $Pr(H|E) > Pr(H)$. Thus, confirmation as increase in firmness compares the conditional probability of H given E with the unconditional probability of H. Notice that these two notions of confirmation do not march in lockstep. Thus, it is possible for the conditional probability of H given E to be larger than the unconditional probability of H without its being larger than any given r, and it is possible for the conditional probability of H given E to be larger than any given $r < 1$ without its being larger than the unconditional probability of H—indeed, it might even be *lower* than the unconditional probability of H.

Which notion is the relevant one? Confirmation as firmness is relevant to the question about the relationship between the degree notion of credence and binary notions such as (full) belief or acceptance. For it may be thought that we are justified in believing a proposition if and only if the rational credence in that proposition given our total evidence is higher than some threshold. This is known as a "Lockean" conception of the relationship between credences and full beliefs. But given that, as we just said, it is possible for $Pr(H|E) > r$ even when $Pr(H|E) < Pr(H)$, confirmation as firmness clearly does not answer the question of when some body of evidence E is evidence *for* a proposition H. Carnap therefore adopts the notion of increase in firmness as the explanation of the idea that E confirms H.

Notice that this notion of confirmation is *comparative* and not *quantitative*—a distinction also introduced by Carnap. If we want to explain the quantitative notion of *how much* some evidence confirms a hypothesis, we need to introduce a *measure* of confirmation. A natural measure is the "difference measure," where we define the degree to which E confirms H relative to Pr as $Pr(H|E) - Pr(H)$.[10]

As we said, Carnap was ultimately interested not in the notion of confirmation relative to a probability function, but in the absolute notion of confirmation. Another way of stating this is that Carnap was interested in delineating a *unique* function Cr_u with respect to which he could define the confirmation of a hypothesis H by a body of evidence E as $Cr_u(H|E) > Cr_u(H)$. Carnap considered two candidates for Cr_u. The first one, $c^{\dagger}$, is the one mentioned at the end of the previous section.[11] Suppose, again, that we are interested in the color of emeralds. More precisely, suppose that our language contains only the predicate "Green," and that we are interested in whether 1,000 emeralds to be found in a bag are green. Thus, there are 1,000 atomic propositions, which we can represent as G_i, where $1 \geq i \geq 1,000$. There are also $2^{1,000}$ (i.e., in the order of 10^{301})

[10] The distinction between confirmation as firmness and confirmation as degree of firmness appears in the preface to the second edition of Carnap (1950). See Fitelson (1999) for a detailed examination of different measures of confirmation. The difference measure, though natural, has some problems.

[11] Carnap actually distinguishes unconditional probability distributions from conditional ones, whereas I give the same name to both.

state-descriptions. The function $c^{\dagger}$ assigns the same probability $(\frac{1}{2^{1,000}})$ to each one of these state-descriptions. If $c^{\dagger}$ is the function that correctly describes our credences, this means that, in the absence of any evidence regarding the emeralds in the bag, we consider each possible combination of green and non-green emeralds equally likely. Notice that each atomic proposition G_1 will be true in exactly half of the state-descriptions, which means that $c^{\dagger}$ assigns probability ½ to each atomic proposition.

The function $c^{\dagger}$ represents a very natural application of the Principle of Indifference. However, as Carnap himself saw, it cannot be right. For notice that $c^{\dagger}(-|G_i)$, i.e., the conditional probability of any proposition given any atomic proposition, according to $c^{\dagger}$, preserves the probabilities of all the other atomic propositions (which, as we saw, is ½). This is because, for any atomic proposition G_i, we can partition the set of state-descriptions into two cells, one containing only state-descriptions where G_i is true and the other containing only state-descriptions where G_i is false. Moreover, each member of the first cell will correspond to exactly one member of the second cell, differing from the first one only in the truth-value of G_i. Conditionalizing on G_i can then be seen as simply zeroing out half of the state-descriptions (those where G_i is false) and then re-normalizing over the remaining ones. Any atomic proposition will be true exactly in half of the remaining state-descriptions, and given that each state-description will still be assigned the same probability by $c^{\dagger}$, every atomic proposition will still be assigned ½.

If we adopt $c^{\dagger}$ as our ur-credence function, this means that, no matter how many green emeralds we see, our credence that the next one is also green will only be ½. As Carnap put it, this "would be tantamount to the principle never to let our past experiences influence our expectations for the future."[12]

Notice the nature of this objection to (this application of) the Principle of Indifference: we are saying that the principle is incorrect because it conflicts with our intuitions about which inductive inferences (encoded in the conditional probabilities) are correct. There is a conflict here because of the relationship between conditional and unconditional probabilities noticed earlier. The Principle of Indifference determines the unconditional probabilities of every state-description, and these, taken together, determine all the unconditional probabilities. Moreover, if we fix all the conditional probabilities of every proposition given any other one, we thereby determine the unconditional probability of every proposition. Unconditional probabilities and conditional probabilities do not float free from each other. This might seem obvious enough given the definition of conditional probabilities as a ratio of unconditional probabilities, but the consequences of the tight relationship between them are deep.

[12] Carnap (1950), p. 565.

The second function Carnap considered, c^*, results from a different application of the Principle of Indifference. To understand this different function, we need to first introduce Carnap's notion of a *structure description*. Some state-descriptions are *permutations* of other state-descriptions: they differ only in which individuals have which properties. Thus, in our example, the state-description that assigns G to every even-numbered individual and $\neg G$ to every odd-numbered individual and the state-description that assigns $\neg G$ to every even-numbered individual and G to every odd-numbered individual are permutations of each other. In general, two state-descriptions are permutations of each other if and only if one can be obtained from the other by permutation of individuals. In our example, two state-descriptions are permutations of each other if and only if they "say" that the same *number* of emeralds are green—regardless of which ones it classifies as green and which ones as non-green. A structure-description, then, is just a disjunction of state-descriptions where each disjunct is a permutation of every other disjunct. Thus, in our example there are 1,001 structure-descriptions. The function c^* assigns equal probability not to each state-description, but to every structure-description, and also equal probability to each state-description within each structure-description. Thus, whereas $c^\dagger$ assigns $\frac{1}{2^{1000}}$ to the state-description that says that only emerald number 1 is green, c^* assigns probability $\frac{1}{1{,}001}$ to the structure-description of which that state-description is a disjunct—and because there are 1,000 such disjuncts, it assigns $\frac{1}{1{,}001{,}000}$ to the state-description in question. Thus, c^* considers the possibility where only emerald number 1 is green vastly more likely than $c^\dagger$ does.

The function c^* doesn't have the problem that $c^\dagger$ has. Suppose, first, that instead of our 1,000 emeralds we are considering just two. There are then four state-descriptions but only three structure-descriptions. Applying our previous reasoning to this case, $c^\dagger(G_2) = \frac{1}{2} = c^\dagger(G_2|G_1)$. By contrast, $c^*(G_2) = \frac{1}{2} \neq c^*(G_2|G_1) = \frac{2}{3}$. The same effect translates to our case with 1,000 emeralds. Thus, c^* *does* allow learning from our past experiences.

However, c^* still has problems with the conditional probabilities it gives rise to—for c^* is insensitive to analogical effects of evidence. To see this we need to consider applications where there is more than one property. For instance, the conditional probability that emerald number 2 is green given that emerald 1 is a small green emerald and emerald 2 is small is no larger than the unconditional probability that emerald 2 is green.

Carnap's reaction to this problem was to consider not just one but a continuum of probability distributions of which both $c^\dagger$ and c^* are special cases. In Carnap's later work,[13] this continuum depends on three parameters, which can be set in ways that allow both for learning from experience as well as for certain analogical

[13] Carnap (1971), Carnap (1980).

effects. However, a more fundamental problem affects Carnap's program, one to which no amount of technical sophistication seems to give an answer.

3.7 Grue

Notice that Carnap's search for a unique ur-prior was constrained in a particular way: his method was purely syntactical in that which properties are in question is irrelevant to the assignment of probabilities. A famous problem by Goodman (1979), however, strongly suggests that no such method can yield the correct ur-prior.

Goodman defined a predicate, "grue," as follows: a thing is grue if and only if either it is green and was first observed before time t, or it is blue. To be more precise, this defines a predicate-schema, from where we can get different definitions by instantiating t to different times. Let us, suppose, for instance, that we instantiate t to January 21, 2040.

Now suppose that we are objective Bayesians à la Carnap—we think there is a unique ur-prior which can be syntactically specified, and we think that rational credences evolve by Conditionalization. Let us also suppose that we have had no evidence whatsoever so far (or, only a bit less unrealistically, no evidence whatsoever regarding the color of emeralds), and so that the rational credence function for us just is the ur-prior. Suppose now that we observe thousands of emeralds and find them all to be green. That is to say, our evidence includes propositions of the form G_i, for a bunch of is. This, it seems, should greatly increase our confidence in the proposition that all emeralds are green. That is to say, the ur-prior Cr_u should be such that $Cr_u(\forall x G(x) | G_1 \wedge \ldots \wedge G_i) > Cr_u(\forall x G(x))$.

Goodman's crucial observation is that our evidence also includes propositions to the effect that thousands of emeralds are green and observed before January 21, 2040. In fact, this more detailed description of what our evidence consists in may seem preferable to the previous one, given Carnap's own insistence on the principle of total evidence—that we should always consider the totality of our evidence in figuring out which probability to assign to hypotheses. Of course, whether we describe the evidence as "emeralds 1 through i are green" or as "emeralds 1 through i are green and observed before January 21, 2040" doesn't make a difference as to whether conditionalizing on it should increase our confidence that all emeralds are green. But the more detailed description of our evidence can be summarized as follows: we have observed a bunch of grue emeralds. But, of course, this should *not* greatly increase our confidence in the proposition that all emeralds are grue.

There is a standard Bayesian approach to the grue problem, whose general outlines are as follows. First, it cannot be denied that finding out that a bunch of emeralds are grue should increase our confidence that all emeralds are grue *to some extent*. This is because the hypothesis that all emeralds are grue entails the evidence that emeralds 1 through i are grue, and if H entails E then $Pr(H|E) > Pr(H)$ for any probability function Pr.[14] But what we *can* hold is that the evidence in question confirms the hypothesis that all emeralds are green *much more* than it does the hypothesis that all emeralds are grue. As Carnap would have put it, the answer to the classificatory question of whether a bunch of grue emeralds confirms the hypothesis that all emeralds are grue must be "Yes," but that leaves open the answer to the comparative question of whether the same evidence confirms the green hypothesis more than the grue hypothesis—and also leaves open the quantitative question of to what degree the evidence confirms each of the hypotheses.

Let us tackle the comparative question. Usually, the way this question is examined in the Bayesian literature is by deriving an answer to it from an answer to the quantitative questions of how much the evidence confirms each of the hypotheses. One problem here, already known to Carnap and recently emphasized by Fitelson (see Fitelson (1999)), is that different measures used in the literature are not even ordinally equivalent, and that threatens the robustness of the results appealed to. But in order to see how the question might be tackled in a Carnapian framework we do not need to enter into any of these complications. For, in that framework, only syntactic differences can yield confirmational differences. There is, of course, a difference between "green" and "grue" when "grue" is given its usual disjunctive definition, but that difference is an accident of the language we used to describe the different properties. "Grue" is disjunctively defined in terms of green and blue, but we can also disjunctively define green and blue in terms of grue and bleen (where a thing is bleen if and only if it is blue and first observed before January 21, 2040, or otherwise green). Given that the Carnapian measures only consider the syntactic features of predicates, there is no way for them to distinguish whether a predicate G refers to a property like green or one like grue. Therefore, although a Carnapian measure can distinguish between

[14] Proof: By Bayes's Theorem, $Pr(H|E) = \frac{Pr(E|H)Pr(H)}{Pr(E)}$. By the law of total probability, the denominator in that fraction can be rewritten as $Pr(E|H)Pr(H) + Pr(E|\neg H)Pr(\neg H)$. Given that H entails E, the numerator is 1 and the first summand in the denominator is $Pr(H)$. So, $Pr(H|E) = \frac{Pr(H)}{Pr(H)+Pr(E|\neg H)Pr(\neg H)}$. So we are dividing *Pr(H)*, a number between 0 and 1, by another number guaranteed to be between *Pr(H)* and 1. If the denominator is 1, then $Pr(H|E) = Pr(H)$, but the denominator can be 1 only when $Pr(E) = 1$, which we are assuming is not the case. Therefore, $Pr(H|E) > Pr(H)$. Notice also that in our case the hypothesis entails the evidence because we are sampling from known emeralds—i.e., we assume that all the *i*s are emeralds, and we are just interested in their color. If we were sampling objects from the universe at random, then the evidence would be that a given object is both an emerald and green (and examined before January 21, 2040). *This* evidence is not entailed by either of the hypotheses: the hypothesis that all emeralds are green entails the conditional that if the object is an emerald then it is green, but it doesn't entail the conjunction that the object is green and an emerald (and analogously for the hypothesis that all emeralds are grue).

the different properties, it can only do so via their representation as predicates in a language. And given that we can treat either of the pair "green" and "grue" as primitive and define the other, whether some evidence confirms the green hypothesis more than the grue hypothesis according to a Carnapian measure depends on the language used. But Goodman's point is that a bunch of green emeralds confirms the green hypothesis more than it does the grue hypothesis regardless of the language we choose to describe the hypotheses. This a Carnapian measure cannot do.

3.8 Whither Objective Bayesianism?

Many philosophers took the failure of Carnap's project to mean that, in general, Objective Bayesianism is doomed. But Carnap was fighting with one hand tied to his back: why think that the failure of determining the shape of the ur-prior purely syntactically means that there is no such unique ur-prior? What is wrong, to put it bluntly, with the idea that there is a unique ur-prior, but it is sensitive to the properties over which it is defined (and not just their linguistic expression)? Many philosophers think that the idea is preposterous. In this section I examine two reasons for thinking Objective Bayesianism preposterous: the claim that it goes against a very plausible kind of epistemic permissivism, and the claim that it requires positing unexplainable cognitive faculties.

3.8.1 Objective Bayesianism and Permissivism

Epistemic permissivism is the view that more than one doxastic attitude towards a proposition can be rational, even fixing the evidence and everything else that might matter to the strength of epistemic position the subject has with respect to that proposition. Objective Bayesianism is, of course, not permissive: given an evidential corpus, the ur-prior fixes the unique rational credence towards any proposition. There is a lively dispute about this so-called "uniqueness thesis" in epistemology.[15] But it is interesting to note that one prominent objector to uniqueness, Thomas Kelly, appeals to the allegedly obvious falsity of uniqueness in the fine-grained case as something of a premise for his argument against the thesis in the coarse-grained case. He says:

[15] Richard Feldman formulated and defended the thesis within the framework of coarse-grained epistemology—Feldman (2003), Feldman (2007). Roger White defended it within the framework of Bayesian epistemology—White (2005). Kelly (2010) attacked both theses—see also Rosen (2001) and Schoenfield (2014).

> The Uniqueness Thesis seems most plausible when we think of belief in a maximally coarse-grained way, as an all-or-nothing matter. On the other hand, as we think of belief in an increasingly fine-grained way, the more counter-intuitive it seems. (Kelly (2010), p. 120)

A little later, he adds:

> Although the Uniqueness Thesis is inconsistent with many popular views in epistemology and the philosophy of science, its extreme character is perhaps best appreciated in a Bayesian framework. In Bayesian terms, the Uniqueness Thesis is equivalent to the suggestion that there is some single prior probability distribution that it is rational for one to have, any slight deviation from which already constitutes a departure from perfect rationality. This contrasts most strongly with so-called orthodox Bayesianism, according to which any prior probability distribution is reasonable so long as it is probabilistically coherent. Of course, many Bayesians think that orthodoxy is in this respect overly permissive. But notably, even Bayesians who are considered Hard Liners for holding that there are substantive constraints on rational prior probability distributions other than mere probabilistic coherence typically want nothing to do with the suggestion there is some uniquely rational distribution. With respect to this long-running debate, then, commitment to the Uniqueness Thesis yields a view that would be considered by many to be beyond the pale, too Hard Line even for the taste of most Hard Liners themselves. (Kelly (2010), pp. 120–1)

Kelly's orthodox Bayesians are my Subjective Bayesians, and his Hard Liners move from the Subjective Bayesian extreme to the side of Objective Bayesianism. But Kelly is saying that even Hard Liners are loath to move all the way towards the other extreme. Meanwhile, until the recent explosion of literature on the issue, Uniqueness was the received view in coarse-grained epistemology, so entrenched that it wasn't even explicitly stated as such. Why the difference?

My hypothesis is that the difference has less to do with Uniqueness itself than with other features of Bayesian epistemology. In particular, it is the problem of "false precision"[16] that is making people nervous about Uniqueness in a Bayesian framework. It is one thing to hold that, given certain evidence, there is a unique attitude it is rational to adopt with respect to a certain proposition: believe it, disbelieve it, or suspend judgment on it; it is quite a different thing to hold that, given certain evidence, it is rational to assign a precise point value to a proposition, and not one that differs from it only to an arbitrarily small degree. The fact that there are uncountably many values to pick from, and the related fact

[16] The phrase is from Kaplan (1996).

that there are values that are arbitrarily close to each other, make philosophers uncomfortable in sticking to Uniqueness. But if this is the real reason why some philosophers shy away from Uniqueness, it is not a good one—at least if we take the approach to the Bayesian framework that I suggested in the last chapter.

There, I argued that Bayesianism is literally false. For one, it might be perfectly rational for us to assign a credence less than 1 to a tautology. But that doesn't mean that the Bayesian framework is useless. On the contrary, it is a very useful idealization. Now, in saying that Bayesianism is a useful idealization I do not mean the same thing as when some philosophers say it. David Christensen, for instance, means that Bayesian requirements are requirements of ideal rationality, and that insofar as we violate them we are not ideally rational.[17] That is not what I mean. What I mean is that Bayesianism can help us isolate one aspect of what matters to rational action: our rational partial ignorance about matters of fact. Of course, we are also ignorant, and sometimes rationally so, about logic, but there is a way in which our rational ignorance about non-logical facts contributes to what is rational for us to do, and we cannot isolate that particular contribution if we at the same time try to model our logical ignorance. Thus, we assume logical omniscience not because we see it as an ideal of rationality, but because it makes our lack of factual omniscience easier to study.

I propose that we tackle the issue of false precision in a similar way. Of course, our evidence will in general not require us to assign a point value to propositions, as opposed to a different but arbitrarily close point value. And this is not a regrettable consequence of our cognitive limitations that we can brush off by saying that if only we were ideally rational our evidence *would* require us to assign such point values. Rather, it is in the nature of (most of) our evidence that it does not rationalize one particular point value. Some philosophers would want to model this as well, opting for sets of probability functions rather than a single one.[18] That is one way to go, but another (to be adopted here) is to simply treat the point values as idealizations in the same way that we treat logical omniscience as an idealization: not as an *ideal*, but as an abstraction from a real feature of our rationality that allows us to concentrate on other features.

It is clear that nothing I've said in this section amounts to an argument against permissivism. Rather, I have warned against a *bad* reason for being a permissivist when thinking about epistemology in the Bayesian framework. If your reason for being a permissivist in that context is that you find it unbelievable that evidence might fix a point-valued credence in a proposition (as opposed to an arbitrarily close one), then you are right: it is unbelievable. But you don't have to believe it. If your reason for being a permissivist is not that one, however, I cannot help you.

[17] See Christensen (2004), but cf. Christensen (2007b).
[18] See Kaplan (1996) and Sturgeon (2008), but cf. Elga (2010) and White (2009).

3.8.2 Is Objective Bayesianism Committed to Magic?

If I am right, then, the issue of false precision is one of the factors that makes philosophers uncomfortable about Objective Bayesianism—and, in particular, about a kind of Objective Bayesianism freed from Carnap's syntactic strictures. But maybe there are additional reasons to be skeptical of this kind of Objective Bayesianism.

Recently, Michael Titelbaum has generalized Goodman's argument against the Carnapian project (Titelbaum (2010)). According to Titelbaum, the kind of language sensitivity exhibited by Goodman's grue problem affects not only theories that seek to determine the ur-prior in purely formal terms (in Carnap's case, syntactically), but a much wider class of theories.

Titelbaum poses the problem in terms of an inconsistent triad of propositions:

1. There is an evidential favoring relation that holds in at least some cases in which the evidence and the hypothesis are logically independent.
2. In at least some of these logically independent cases, agents can determine that the body of evidence favors the one hypothesis over the other.
3. Hypotheses about which properties are special are empirical hypotheses that must be determined from an agent's evidence and are logically independent of that evidence. (Titelbaum (2010), p. 478, renumbered)

Titelbaum argues that, in order for an agent to be able to determine when a body of evidence supports one hypothesis over another, she must be able to determine that particular properties are special. We have already made Titelbaum's point for the case of grue. In order to distinguish between the green and the grue hypothesis, a measure of confirmation must "play favorites" (as Titelbaum puts it) with the properties themselves, independently of the language used to describe the hypotheses.

So, if 1 is true and the ur-prior is such that, for example, $Cr_u(Green|E) > Cr_u(Grue|E)$ (where E is the evidence mentioned in the previous paragraph, *Green* is the hypothesis that all emeralds are green, and *Grue* is the hypothesis that all emeralds are grue), then the ur-prior discriminates between these hypotheses even in the absence of any evidence, for $Cr_u(Green) \neq Cr_u(Grue)$. This is a well-known feature of the Bayesian treatment of Goodman's grue problem. In Titelbaum's terms, the evidence can favor the hypothesis that all emeralds are green over the hypothesis that all emeralds are grue only if the ur-prior treats the property of being green as "special."

Titelbaum adds his 3, which, applied to our example, is the claim that whether being green or being grue are special is an empirical matter, and, as such, can only be determined by appeal to evidence. A contradiction then easily follows. If, as 2 holds, the agent can determine that the evidence favors the green over the grue hypothesis even before receiving that evidence, then that means that she can determine that being green is a special property in the absence of evidence. But that directly contradicts 3. Titelbaum concludes that we should give up 1. There is no objective evidential favoring relation with the property attributed to it by 1.

But there is an alternative: to give up 3. To do so we must hold that we can determine a priori, in the absence of any empirical evidence, that certain properties (like green) are special. Titelbaum is of course aware of this possibility, and he argues against it. His main argument consists in pointing out how strange this faculty of ours of being able to determine the specialness of properties a priori would have to be. By which magic, one can hear Titelbaum asking, can we determine a priori that green, and not grue, is special?

There is no magic. An answer that Titelbaum doesn't consider is that we determine the specialness of green by determining that the evidence in question would support the green over the grue hypothesis. It's not that we consider the property of being green and the property of being grue in isolation, and then detect with a specialness-detecting faculty that green is special and grue is not. Rather, we consider Goodman's case and we determine that the evidence in that case supports the green hypothesis over the grue hypothesis, and that this would be so even if we are unfortunate enough to live in a world where all emeralds are grue. We then reflect a bit on Bayes' Theorem and realize that the evidence can have this differential effect only if the unconditional probability of the green hypothesis is higher than that of the grue hypothesis.

It is worth noting some differences between this answer and one that appeals to the claim that whereas *green* is a natural property, *grue* isn't—or to the comparative claim that *green* is more natural than *grue*. This latter claim is central to David Armstrong's and David Lewis's metaphysics. Armstrong (1978) posited the existence of universals as a solution to the "one over many" problem. Lewis (1983) distinguished between universals as Armstrong conceived of them and properties. Properties are simply sets (or classes, if too big to be a set). Take any things you want: there is a property they have in common, for there is a set that they all belong to. Universals are different: they underlie *real* similarity relations. Whereas any two things share a property *just in virtue of* belonging to a same set, any two things have something in common only if (and because) they share a universal. Properties are abundant, universals are sparse. In particular, there is a property *being grue*, but there is no universal *being grue*. We may, if we want, say that grue things have something in common, but that is true only in the deflationary sense that there is a set to which they belong. Green things, by contrast, really do have something in common: they share the universal *green*.

Lewis puts this distinction between properties and universals to work in a variety of ways. Of particular interest to us now is the work to which he puts it in the philosophy of language and mind. Lewis's solution to Putnam's model-theoretic argument (see, for instance, Putnam (1980)), in particular, consists in claiming that the reference of our terms is fixed not just by our intentions (reflected in our use of the term), but also by the *eligibility* of the different candidates. The more natural a property, for instance, the more eligible it is to be the referent of one of our predicates. Lewis applies the same idea to the interpretation of thought as well as language.

Lewis may help Carnap out of the problem Goodman posed. The problem, recall, was that there is no deep syntactical difference between the predicates *green* and *grue*, and thus that there is no way for an ur-prior determined in purely syntactical terms to distinguish between them—and, thus, to explain why observing green emeralds should raise our credence that all emeralds are green more than it does our credence that all emeralds are grue. But if greenness is more natural than grueness, then a language that treats "green" as a primitive and disjunctively defines "grue" will carve nature more closely to its joints than a language that treats "grue" as a primitive and disjunctively defines "green." Therefore, the metaphysics itself will give us a reason to prefer one language over the other, and then the Carnapian measure will work with a language that syntactically distinguishes between "green" and "grue."

This is not the solution I am proposing. I am not saying that there is a metaphysical difference between greenness and grueness. I am saying that there is an epistemic difference. And, indeed, there must be an epistemic difference for Goodman's problem to even get off the ground. We readily agree with Goodman that a bunch of green emeralds must confirm the green hypothesis more than the grue hypothesis (even if we don't go as far as to claim they must not confirm the grue hypothesis at all). So the very posing of the problem shows that we have access to that kind of epistemic fact. There need not be anything special about greenness itself, as opposed to grueness. My epistemic point is compatible with Lewis's metaphysics of degrees of naturalness among properties, but is also compatible with the denial of that view.

Of course, Titelbaum can say that there is still something mysterious in the procedure whereby we determine that the evidence must confirm the green hypothesis to a greater degree than it confirms the grue hypothesis. True, we need not have any weird faculty whose job is to determine the specialness of properties, but now we are appealing to a faculty that allows us to detect when some evidence favors a hypothesis over another. And the nature and functioning of that faculty, Titelbaum can argue, is just as strange as the nature and functioning of a faculty for directly detecting specialness of properties.

Maybe there is something mysterious about a priori knowledge in general, and perhaps even something more mysterious still about our a priori capacity to figure out evidential relations. But the shock value of holding that we can figure out evidential relations a priori is considerably less than that of holding that we can figure out which properties are special (in Titelbaum's sense) a priori. Some philosophers have even pointed to this capacity of ours as the source of the clearest example of so-called deeply contingent a priori knowledge. For instance, some philosophers have held that we can know a priori that if the evidence in question is true, then all emeralds are green, even though this conditional is as contingent a proposition as they come (see Hawthorne (2002)).

3.9 Conditionalization and Ur-Prior Conditionalization

We have been looking at the difference between Subjective and Objective Bayesianism through the prism of ur-priors. We know that if a subject's credences are probabilistically coherent and updated by Conditionalization, then there will be uncountably many ur-priors such that the credences of that subject at any given time are the result of conditionalizing any of those priors on the evidence the subject has at that time. Subjective Bayesianism is the thesis that this is all that theoretical rationality requires of subjects. Objective Bayesianism, by contrast, requires in addition that the credences of a subject at any given time be obtainable by conditionalizing on the uniquely correct ur-prior. Notice that the multiplicity of ur-priors for the Subjective Bayesian has two sources. First, if the subject in question does not receive every possible body of evidence throughout her life, then there will be, as I just said, uncountably many ur-priors with the requisite property. If we can somehow fix how the subject's credences would have evolved in answer to every possible body of evidence, this source of multiplicity of ur-priors would vanish, and there will be a unique ur-prior with the requisite property. But, even in this case, there is still a difference between Subjective and Objective Bayesianism, because the Subjective Bayesian will say that the identity of this ur-prior doesn't matter: as long as there is one such ur-prior (and it is guaranteed that there will be one if the credences evolve by Conditionalization), then it doesn't matter, for instance, whether it assigns a higher prior to the green hypothesis or to the grue hypothesis. The Objective Bayesian, by contrast, holds that a subject's credences will be rational only if they result by Conditionalization from one specific ur-prior.

Let us now take a closer look at the requirement of Conditionalization. One problem with such a requirement results from what in chapter 2 we called the stickiness of extreme probabilities under Conditionalization. This means that if a subject whose credences evolve by Conditionalization ever assigns credence 1 or 0 to a proposition, then that proposition will never receive any other credence. For suppose that there is a proposition P such that $Cr_t(P) = 1$. Now suppose that the subject updates her credences by Conditionalization, so that $Cr_{t'}(P) = Cr_t(P|E)$, where E is the new evidence the subject received between t and t'.

Then, by Bayes's Theorem, $Cr_{t'}(P) = \frac{Cr_t(E|P)Cr_t(P)}{Cr_t(E)}$. But, given that $Cr_t(P) = 1, Cr_t(E|P) = Cr_t(E)$.[19] So, $Cr_{t'}(P) = \frac{Cr_t(E)Cr_t(P)}{Cr_t(E)} = Cr_t(P) = 1$. A similar argument shows that credence 0 is also sticky in this sense.

Thus, the Bayesian requirement of Conditionalization ensures the cumulative nature of evidence, in the sense that once a proposition becomes part of the subject's evidence, it acquires credence 1 and never loses it. But evidence is plainly not cumulative in this sense. Forgetting is not a failure of rationality, not even of perfect rationality. Thus, although my degree of belief in the proposition that I am typing right now may well be part of my evidence at this moment, it will certainly not be some time from now.[20] Moreover, Bayesian Conditionalization is not just incompatible with forgetting, but even with the *threat* of forgetting. Frank Arntzenius's Shangri-La case illustrates this (Arntzenius (2003)). There are two paths to Shangri-La, the Path by the Mountains and the Path by the Sea. The guardians will toss a fair coin. If it lands heads, then you will take the Path by the Mountains and you will retain your memory of having done so. If it lands tails, you will take the Path by the Sea, but upon arriving at Shangri-La your memory of the journey will be erased and replaced by a fake memory of having taken the Path by the Mountains. Before the coin is tossed, you assign credence ½ to the proposition that it lands heads. Suppose that it does indeed land heads. Then, as you are traveling through the Path by the Mountains, your credence in heads increases to 1. Upon reaching Shangri-La, however, you rationally lower your credence in heads to ½ again—and this despite the fact that you forgot nothing. The mere threat of forgetting is therefore incompatible with Bayesian Conditionalization.

Moreover, the stickiness of extreme probabilities under Bayesian Conditionalization is incompatible not just with memory loss or its threat, but also with the phenomenon of "knowing less by knowing more" facilitated by the existence of misleading evidence. Here's an example from Williamson (2000). Suppose that I show you one black and one white ball, and I put them in an urn. Let us say that it is now part of your evidence that the urn contains one black and one white ball. Now you take one ball at random out of the urn, notice its color, and put it back in. You repeat this experiment 10,000 times, and every time the ball is black. If you were to respect Bayesian Conditionalization, you would have to remain certain that there is a white ball in the urn. The rational thing to do, however, is to

[19] There are different ways of proving that result. Here's one: by the law of total probability, $Cr_t(E) = Cr_t(E|P)Cr_t(P) + Cr_t(E|\neg P)Cr_t(\neg P)$. Given that $Cr_t(P) = 1$, $Cr_t(\neg P) = 0$, and so the second summand in that application of the law of total probability is 0, and so $Cr_t(E) = Cr_t(E|P)$.

[20] You may think that my writing down the sentence will make it remain in my evidence, thus rendering the case ineffective as a counterexample to Bayesian Conditionalization. Notice, however, that I referred to the proposition in question indexically, thus saving the counterexample. And no, I will not tell you when I wrote that sentence. I already forgot.

suspect that your eyes played a trick on you earlier or that something else went wrong and there isn't any white ball on the urn.

These problems with Conditionalization can be handled by changing that requirement to the requirement of Ur-prior Conditionalization:

Ur-prior Conditionalization: $Cr_t(P) = Cr_u(P|E)$,

where Cr_t is the subject's rational credence function at t, Cr_u is the subject's ur-prior credence function (which will be the same for every subject according to Objective Bayesianism), and E is all the evidence the subject has at t. The difference between Bayesian Conditionalization and Ur-prior Conditionalization can be seen by applying both to the case we just talked about. As we just saw, according to Bayesian Conditionalization you should assign credence 1 to the proposition that there is a white ball in the urn even after receiving what we would pre-theoretically describe as massive amounts of evidence against that proposition. Ur-prior Conditionalization works differently. At t_1, after seeing the balls being put into the urn but before performing the experiment, it is part of your evidence that there is a white ball on the urn, and so you conditionalize Cr_u on this proposition, with the result that you assign it credence 1. At time t_2, after you perform the experiment, it is no longer part of your evidence that there is a white ball in the urn, so you conditionalize Cr_u on the results of the experiment, with the consequence that now you assign a much lower credence to the proposition that there is a white ball in the urn. Let the evidence you have at t_1 be E_1, and the evidence you have at t_2 be E_2. According to Bayesian Conditionalization, $Cr_{t_2}(P) = Cr_{t_1}(P|E_2)$, and thus $Cr_{t2}(E_1) = Cr_{t1}(E_1) = 1$, whereas according to Ur-prior Conditionalization $Cr_{t_2}(P) = Cr_u(P|E_2)$, which allows for $Cr_{t2}(E_1) < 1$.

Bayesian Conditionalization forces evidence to be cumulative, whereas Ur-prior Conditionalization doesn't. Both views about credence updating treat the evidence a subject has at a time as an exogenous input to the update procedure. But whereas Bayesian Conditionalization guarantees that once some proposition has been provided as input it will remain as such forever, Ur-prior Conditionalization imposes no such artificial guarantee. In any case, however, both views highlight another important aspect with respect to which all the Bayesian views so far presented are radically incomplete: none of them contains a theory of evidence. I now turn to that issue.

3.10 Knowledge-First Epistemology and E = K

Timothy Williamson's *Knowledge and Its Limits* is widely and rightly regarded as one of the central books in epistemology of the last few decades. In that book, Williamson delineates, argues for, and fills in the program of "knowledge-first" epistemology. That program has many components: the claim that the post-Gettier project of defining "knowledge" in terms that do not themselves presuppose the notion of knowledge is doomed to failure; the claim that knowledge is a bona fide mental state; the claim that knowledge is metaphysically "prime" (metaphysically unanalyzable); the claim that knowledge itself can be used as the main primitive in epistemological theorizing. It is this last claim that I am interested in here. In particular, I am interested in Williamson's identification of the evidence a subject has at a time with the propositions the subject knows at that time (which he summarizes in the equation E = K, and which we called "Factualism"). Thus, Williamson endorses Objective Bayesianism and supplements it with an account of evidence—something that is sorely needed in any Bayesian account.

It is worth noting here that Williamson himself does not think that his evidential probabilities can be identified with the credences a rational subject would assign given the specified evidence. His reasons for denying this, however, have to do with the fact that a rational subject, according to Williamson, would not have the evidence we have. This is because a rational subject would, for instance, recognize T^* as a tautology, and so assign it credence 1. I agree that which credences it is rational for us to assign are not those that we would assign if we had no computational limitations whatsoever, but I disagree with the claim that the ideally rational subject is one without computational limitations. In any case, the proposed identification is of evidential probabilities with the credences a rational subject would assign *given the specified evidence*, not given the evidence the rational subject would have if he were in our situation. Moreover, in line with my discussion in chapter 1 of the problem of logical omniscience, I am bracketing off computational limitations unless ignoring them affects some result I am interested in.

Although Factualism is obviously a claim that is congenial to the general project of knowledge-first epistemology, it is not clear that Williamson's Objective Bayesianism as a whole is so congenial. For his Objective Bayesianism includes a commitment to the uniqueness of the rational ur-prior. This ur-prior determines two related things: which credences it is rational to assign to any proposition absent any evidence, and which credences it is rational to assign to any proposition given any body of evidence. Factualism guarantees a central role for knowledge to play in the overall epistemology, but a commitment to the uniqueness of the ur-prior guarantees a central role for a knowledge-independent notion of rationality as well.

3.11 Conclusion

Regardless of how congenial to knowledge-first epistemology Williamson's overall package is, it does at least provide us with a complete Objective Bayesian theory that we can assess. That will be the task of the next two chapters, where I will argue against Factualism by way of arguing against the kind of knowledge-based decision theory to which it gives rise. A prominent alternative to Factualism as a theory of evidence is provided by the view according to which our experiences themselves are our evidence (Williamson calls this the "phenomenal conception of evidence," and, following Dancy, I called it "Psychologism"). I disagree with this view as well. There is a better version of Objection Bayesianism, I will argue, that rejects Factualism but that doesn't revert to Psychologism.

4

Knowledge-Based Decision Theory?

4.1 Introduction

In the last chapter I presented a version of Objective Bayesianism which embeds Williamson's Factualism as a theory of evidence. The view is that the credences that are rational for a subject to assign to any proposition p at a time t are given by $Cr_u(p|E_t)$, where Cr_u is the ur-prior and E_t is the conjunction of all the propositions the subject knows at t. My aim in this chapter is to show that this version of Objective Bayesianism runs into trouble when we consider how it interacts with the traditional theory of practical rationality sketched in chapter 2. Because of Factualism, the view has it that false beliefs cannot rationalize actions, but they can. Factualism also runs into trouble for cases of non-basic knowledge—for instance, inferred beliefs should not be counted as part of one's evidence, even when those inferred beliefs amount to (inferential) knowledge.

The theory of practical rationality is concerned with the rationality of actions (or intentions to perform actions).[1] According to a theory familiar to philosophers, the rationality of actions is determined by the beliefs and desires of the subject in question. Thus, according to this Humean theory, and to a first approximation, if Mary believes that her friends are at the bar and wants to meet her friends, it is rational for her to head to the bar.

The are several reasons why that example is only a first approximation to what a Humean theory would say. In the first place, it may be that although Mary wants to meet her friends, she has an even stronger desire to avoid her nemesis, whom she also believes to be at the bar. In the second place, it may be that her degree of belief in the proposition that her friends are at the bar is higher than her degree of belief in the proposition that her nemesis is at the bar. The first complication militates against the idea that it is rational for her to head to the bar, whereas the second makes this idea look good again. Whereas these observations call for modifications to the Humean theory (to be developed in section 4.2), some

[1] The retreat from actions to intentions may be motivated by the thought that, just as there cannot be moral luck, there cannot be rational luck either. Given that whether an intention results in the corresponding action is thought to be out of the direct control of the agent, the proper subject for the theory of practical rationality is then restricted to intentions, over which the agent is supposed to have direct control. As is the case with moral luck, arguably this retreat is not well motivated: neither are the actions outside of the control of the agent in a relevant way nor are his intentions within his complete and direct control either. Thus, in what follows I assume that practical rationality concerns the rationality of actions, although the main arguments of this book do not depend on this assumption.

Being Rational and Being Right. Juan Comesaña, Oxford University Press (2020). © Juan Comesaña.
DOI: 10.1093/oso/9780198847717.001.0001

philosophers are more directly opposed to anything in the general vicinity of a Humean theory. Some of these philosophers, call them "Kantians," think that actual desires (or preferences, or other conative attitudes) of the subjects have approximately no bearing on what it is rational for them to do. Rather, according to one version of Kantianism, what matters is which desires it would be *rational* for these subjects to have. Humeans do distinguish between instrumental and basic desires, and although they do grant that instrumental desires may be qualified as rational or not (according to whether they align in the right way with basic desires), they do not think that it makes sense to think of the basic desires themselves as rational or not. Although my sympathies lie with Humeanism and I will continue to write as if it is right in its opposition to Kantianism, the main arguments of this book will not be affected by this choice.

Another relatively friendly amendment to the Humean picture will play a central role in this chapter. The amendment consists in saying that, although Kantians are wrong in requiring that the desires in question be rational, the beliefs (and credences) in question do need to be rational. Thus, suppose that Mary's belief that her friends are at the bar is simply the result of wishful thinking. Then, although a subject with that combination of beliefs and desires who does not go to the bar may in some sense be inferior to one who does (more on this in chapter 7), it would be strange to say that the rational thing to do for Mary is to go to the bar. I will sometimes refer to this constraint as the view that only justified attitudes can justify. But care is needed to interpret that: I do not mean to say that actions (or intentions, or other attitudes) can only be justified by justified attitudes, but just that, when they are justified by attitudes, the justifying attitudes need to be themselves justified. Because of the prominence of the thesis that practical rationality presupposes theoretical rationality in Fumerton (1990), I will sometimes call it "Fumerton's thesis."

If what I've said so far is right, then in order for Mary's action of heading to the bar to be rational, it is required that her belief that her friends are at the bar itself be rational. But, in the order of discovery, we need not *first* establish the rationality of beliefs to determine the rationality of action. We might well start with the judgment that, in a case as described, the subject's action is rational and then infer from this that the beliefs on which the action is based are also rational. This is the route taken in this chapter. I start from judgments about the rationality of certain actions, and then conclude that the beliefs in question are also rational. Given that the beliefs in question are false, I conclude that there can be false rational beliefs. This is not yet in direct conflict with Factualism, because Factualism requires only that one's evidence be true, not that all of one's justified beliefs be true.[2] But some

[2] Despite some recent arguments from Williamson, which seek to establish (unsuccessfully, in my opinion) the stronger view from the weaker. See Williamson (2013a), Williamson (2013b), and Williamson (forthcoming), and cf. Cohen and Comesaña (2013a), Cohen and Comesaña (2013b), and Cohen and Comesaña (forthcoming).

of the rational beliefs underlying rational actions are non-inferentially acquired, and arguably those beliefs are parts of one's evidence. Moreover, those beliefs will play the role that evidence plays in traditional decision theory: a state needs to be considered for a decision just in case it is compatible with the propositions believed. I will therefore conclude that not only can false beliefs rationalize actions, but false beliefs can rationalize actions because they can be part of one's evidence.

The view that rational action requires rational belief but tolerates false belief is under pressure from two directions: the view that rational action requires mere belief, whether rational or not, and the view that, although rational action requires rational belief, rational beliefs cannot be false. I tackle these two objections in turn, starting with a brief recap on traditional decision theory.

4.2 Traditional Decision Theory and Practical Dilemmas

Traditional decision theory, as presented in chapter 2, is the contemporary descendant of the Humean theory of practical rationality briefly sketched in section 4.1. In place of the Humean theory's beliefs and desires, traditional decision theory posits credences and preferences, and in place of the Humean theory's instrumental reasoning, traditional decision theory posits maximization of expected utility.[3]

Let us recall the example used in chapter 2. There we assumed that the partition of states relevant to your decision is given by whether it rains or not. Your available actions are going to the movies or going to the park. The following lists all the possible outcomes of those actions, the utility that the function that represents your preferences assigns to them, and your credence in each of the possibilities of the partition:

$u(\textit{movies with your friend}) = 7$

$u(\textit{movies alone}) = 1$

$u(\textit{park with your friend}) = 15$

$u(\textit{park alone}) = 6$

$Cr(\textit{rain}) = .7$

$Cr(\textit{no rain}) = .3$

[3] More accurately, as I will argue in section 4.3, traditional decision theory posits credences *in addition to* beliefs, for it implicitly recognizes a role for belief in the notion of evidence.

In table form:

	.7	.3
	It rains	It doesn't rain
Go to the movies	Movies with your friend 7	Movies alone 1
Go to the park	Park alone 6	Park with your friend 15

Given that these are your credences and preferences, traditional decision theory has it that it is rational for you to go to the park.

Traditional decision theory is usually married to Subjective Bayesianism as a theory of rational credences. Recall that according to Subjective Bayesianism, the credences of a subject are rational provided only that they are probabilistically coherent at any given time and that they evolve by Conditionalization on the subject's new evidence (where this notion of evidence is usually left unexplained by Subjective Bayesianism). This combination of traditional decision theory with Subjective Bayesianism therefore clashes with Fumerton's thesis, at least insofar as Subjective Bayesianism is not thought to be a plausible theory of rational credences. In chapter 3 I argued precisely for that conclusion. But there is an argument that Subjective Bayesianism is the right theory of credences to use in decision theory even if it is not the right theory of the rationality of credences. The argument has it that even if Subjective Bayesianism is wrong about the rationality of credences because (say) more than mere probabilistic coherence is required for the credences that a subject has at a time to be rational, still, credences as an input to the expected utility calculation should only be required to be probabilistically coherent.

In the binary realm, the view that rational action does not require rational belief has recently been espoused by Parfit (2011)—but what is there to be said for it? In epistemology, the claim that unjustified beliefs cannot justify lies at the center of the venerable regress argument, and can well be called the traditional view about inferential justification—which doesn't mean that it hasn't been attacked (see, for instance, Wright (2004)). Why would it be any different in the case of practical rationality?[4] Given the framework of traditional decision theory, the justification of action is all inferential, in the following sense: in that framework, an action is rational just in case it maximizes expected utility, and the maximization of expected utility calculation takes as an input the credences of the subject in question (as well as his beliefs, as we shall see). This is analogous to the case

[4] To be fair to Parfit, he does address this very question in the work cited, but his answer strikes me as a simple restating of his view.

where a belief is justified by another belief,[5] and so the analogy suggests that the credences that are the input to the expected utility calculation must themselves also be justified.

But there is an argument for the view that Fumerton's thesis is wrong because all doxastic attitudes, justified or not, can justify. I am not aware of any published version of this argument, but it seems to me to be the motivation behind at least some of the versions of subjective Bayesian decision theory.[6] The argument is that unless we allow that all doxastic attitudes, justified or not, can serve as the inputs for further justified attitudes, we must face dilemmas. The argument goes like this. Suppose that only justified attitudes can justify. Suppose, for instance, that you witness a runaway trolley which cannot be stopped. The trolley is approaching a fork in the tracks, however, and you can pull a switch which will make it go on the right-hand side track. If you do nothing, the trolley will continue on the left-hand side track. You know that there is someone tied to the tracks up ahead, and you want to save them, but you don't know on which track the potential victim is. Someone up ahead screams to you that the potential victim is on the left-hand track. However, you unjustifiedly think that the person screaming at you is the potential victim's nemesis and wants them dead. You therefore form an unjustified high credence in the proposition that the potential victim is on the right-hand side track. If only justified attitudes can justify, then which action are you justified in performing in this situation? You are not justified in not pulling the switch, for you have no justified attitude that justifies that action. But *not* pulling the switch is not justified either, for that action is not justified by any of your actual attitudes.[7] You do not have enough justified attitudes to go around, and so if only justified attitudes can justify, then there is no action (in the broad sense of "action," which includes omissions) that is justified for you here. Hence, you face a dilemma.

I think that the correct answer to this objection is that you do indeed face a dilemma in that situation, but it is not a problematic one.[8] There are three related distinctions which result in three ways of classifying dilemmas, and the dilemmas countenanced by Fumerton's thesis fall on the right side of all three distinctions.

The first distinction that I want to apply to practical dilemmas needs a bit of set up. There is a distinction, familiar to epistemologists, between the justification of a

[5] Strictly speaking, given the view developed in chapter 6, what justifies the target belief is the *content* of another belief, which is possessed as a reason in virtue of the fact that it is the content of a justified belief.

[6] Ellery Eells once told me, in defense of a Subjective Bayesian version of decision theory: "You have to work with what you have to work with," meaning (I think) that there is no alternative to saying that the rational actions are those rationalized by your actual credences (whether they are justified or not). Eells may have had in mind the argument I am about to present.

[7] Notice that there is no third option here consisting in neither pulling nor not pulling the switch. This is an instance of the oft-noticed asymmetry between actions and beliefs, in that there is no analogue of suspension of judgment for the case of action.

[8] I will argue later in chapter 5 that knowledge-based decision theory does face problematic dilemmas.

proposition for a subject, on the one hand, and the justification of the mental state of believing a proposition, on the other. "Belief" is notoriously ambiguous between the state and its content, which obfuscates the distinction. Sometimes the distinction is marked in the terminology of "propositional" vs. "doxastic" justification, but I prefer Goldman's terminology of *ex-ante* vs. *ex-post* justification, for it doesn't suggest that the distinction applies only to doxastic attitudes (see Goldman (1979)). A doxastic attitude towards a proposition is *ex-ante* justified for a subject, regardless of whether the subject adopts the attitude, just in case the subject's epistemic position vis-à-vis that proposition fits that attitude.[9] By contrast, for a subject's belief to be *ex-post* justified the belief not only needs to exist (obviously), but it needs to be properly based.[10]

We can generalize the *ex-ante*/*ex-post* distinction to intentions and (derivatively) to actions. Thus, suppose that given your preferences and *ex-ante* rational credences there is an action *A* which maximizes expected utility for you. We can then say that intending to perform (and, derivatively, performing) *A* is *ex-ante* justified for you. If you do perform *A*, and you do it on the right basis (where presumably this right basis will require that your preferences and credences play the right kind of causal role in motivating you to act), then your performance of *A* counts as *ex-post* justified.

The first distinction that I want to apply to practical dilemmas is, then, the distinction between merely *ex-post* and *ex-ante* dilemmas. Merely *ex-post* practical dilemmas are situations where there is no action that you can rationally perform, although there is an action that is *ex-ante* justified for you to perform. If a practical dilemma is (also) *ex-ante*, then not only is there no action that you can rationally perform, but there is no action that is *ex-ante* rational for you to perform. The practical dilemmas countenanced by Fumerton's thesis are of the merely *ex-post* variety. For go back to the trolley case presented a few paragraphs back. Although you have an unjustified high credence that the potential victim is on the right-hand track, you are *ex-ante* justified in having a high credence that the potential victim is on the left-hand track. Thus, although now that you have adopted an unjustified high credence in the proposition that the victim is on the right-hand track there is no action you can rationally perform, there is an action that it is rational for you to perform, namely to not pull the switch. Unfortunately, you have no relevant *ex-post* justified credences, and so there is no *ex-post* action it is

[9] I use the vague notion of strength of epistemic position in order to remain neutral on first-order epistemological disputes. An evidentialist, for instance, could explain *ex-ante* justification by saying that a doxastic attitude is *ex-ante* justified for a subject provided that it is sufficiently supported by that subject's evidence.

[10] I use the vague notion of proper basing to (once again) remain neutral on first-order issues. An evidentialist, for instance, could explain *ex-post* justification by saying that a belief is *ex-post* justified just in case it is *ex-ante* justified and based on evidence that supports it. More on evidentialism and the basing relation in chapter 9.

rational for you to perform, but not pulling the switch is the action that maximizes utility relative to your preferences and your *ex-ante* justified credences.

The second distinction that I want to apply to practical dilemmas goes back at least to Aquinas, and is the distinction between self-imposed and world-imposed dilemmas. A dilemma is self-imposed if it arises as a consequence of something you (freely) did. Examples will necessarily be controversial, for some philosophers think that there are no real dilemmas of any kind, but if you make two incompatible promises (for instance, you promise to pick up your friend from the airport and to wait for the mail at the same time) then, whatever you do, you break one of those promises—and so, many would say, whatever you do you do something wrong. But the dilemma was self-imposed: you freely made the two promises, and you face the dilemma as a direct consequence of those free actions of yours. By contrast, world-imposed dilemmas arise independently of any free action of the subject. Examples here are even harder to find (I will argue later that knowledge-based decision theory leads to world-imposed dilemmas, and is implausible for doing so), but conceptually the distinction is not hard to grasp. The kind of dilemmas that are a consequence of Fumerton's thesis are self-imposed. I do not want to get here into a discussion about doxastic voluntarism, but even if we grant that the formation of doxastic attitudes is not within your control in the same sense that the formation of your intentions or the performance of your actions is (something that I am not totally convinced we should grant), still the formation of doxastic attitudes is sufficiently "up to you" in the sense required for them to count as something you freely did, in a broad sense of "doing." You are responsible for your attitudes in a way in which you are not responsible for your tics. Thus, in our trolley case, it was up to you to adopt a justified high credence in the proposition that the victim is on the left-hand track, and so the dilemma in question is self-imposed.

Finally, we should distinguish between recoverable and non-recoverable dilemmas. This distinction appeals to the same idea of an action being "up to you" in the broad sense just explained. A dilemma is recoverable just in case there is something you can do, an action that is up to you to perform, such that, if you did it, you would no longer be in a dilemma. A dilemma is non-recoverable just in case there is no such action. Again, examples will be controversial here, particularly because many philosophers will be inclined to think that what I am calling recoverable dilemmas are not real dilemmas at all. I'm not particularly keen on denying that assertion. For the dilemmas envisaged by Fumerton's thesis are recoverable, and so much the better for the view if recoverable dilemmas are not real dilemmas. Those kinds of dilemmas are recoverable because there is something the subject can do—namely, adopt the justified credence—such that, were he to do it, he would no longer be in a dilemma. Thus, if the subject in the trolley case were to adopt a justified high credence in the proposition that the victim is on the left-hand track, then the action which was previously *ex-post* irrational for him to perform—to not pull the switch—will become *ex-post* rational.

I have distinguished between merely *ex-post* and *ex-ante*, world-imposed and self-imposed, and recoverable and non-recoverable dilemmas. The kind of dilemmas that are a consequence of Fumerton's thesis all fall on the right side of these distinctions: they are merely *ex-post*, self-imposed, and recoverable. And while *ex-ante*, world-imposed, and non-recoverable dilemmas might well be problematic, this kind is not. The existence of dilemmas is problematic only insofar as they mean that, through no fault of their own, subjects can find themselves in situations where nothing they can do qualifies as rational (or as moral, in the case of moral dilemmas). If a dilemma is merely *ex-post*, then this means that there is something the subject can do which qualifies as rational—it's just that the subject cannot rationally do it. If the dilemma is recoverable then there is something the subject can do such that, after doing it, he is not even in a merely *ex-post* dilemmatic situation, for now there is an action he can rationally perform. And finally, if the dilemma is self-imposed, then in an obvious sense the "through no fault of their own" clause is not satisfied. Therefore, Fumerton's thesis does not entail the existence of dilemmas in any problematic sense of the term.

4.3 The Role of Beliefs in Traditional Decision Theory

The view defended in this chapter is the combination of Fumerton's thesis (that rational action presupposes rational belief) with the claim that rational action tolerates false belief. In the previous section I defended the first part of that view against the charge that it generates problematic dilemmas. In sections 4.5 and 4.6 I will argue against knowledge-based decision theory, a view that incorporates the kind of knowledge-first Objective Bayesianism described in chapter 3 and according to which rational action does not tolerate false beliefs.

But first I will address a worry that may have already occurred to some: why do I talk here about the justification by "attitudes" generally, and, in particular, why do I talk about beliefs, when the Bayesian framework explained in chapters 2 and 3 does away with such coarse-grained attitudes and deals exclusively with fine-grained credences? The answer is that I do not believe that the Bayesian framework does away with beliefs. Despite appearances, there is a place in traditional decision theory for the binary notion of belief.

Consider, once again, the following example of a decision matrix:

	.7	.3
	It rains	It doesn't rain
Go to the movies	Movies with your friend 7	Movies alone 1
Go to the park	Park alone 6	Park with your friend 15

It might look as if the table represents only the subject's credences in states, the subject's preferences over outcomes, and the actions available to the subject. But there is information represented implicitly by what is *not* on the table. Notice that although the states are labelled "It rains" and "It doesn't rain," they really represent, respectively, the state where it rains and your friend goes to the movies and the state where it doesn't rain and your friend goes to the park. But these are not, of course, all the possible states. There are also states where it rains and your friend goes to the park, and states where it doesn't rain and your friend goes to the movies. For that matter, there are also states where it rains and your friend goes to the movies and aliens abduct you on your way to the movies, etc. Relatedly, notice that the states in question are not whole possible worlds, but rather equivalence classes of possible worlds. For instance, the state where it rains and your friend goes to the movies includes worlds where this happens and the spatial location of all the grains of sand in the Sahara is exactly as in the actual world, but also worlds where they are somewhat differently located. So there are two salient questions about the states that are represented in the matrix: why only these states, and no others, and why states at precisely this level of description, and no finer- or coarser-grained ones?[11]

The question about the fineness of grain of the states has at least a clear normative answer: once a state is represented in the matrix, it should be as finely individuated as it is needed to represent worlds that make a difference to the expected utility calculation. For instance, in the case at hand we distinguish between worlds where it rains and worlds where it doesn't because going to the movies when it rains has a different utility than going to the movies when it doesn't rain. By contrast, the precise location of grains of sand in the Sahara is not represented because the utilities are not sensitive to small enough differences in their location.[12]

The difference between a possible world where the grains of sand in the Sahara have their actual spatial location and one where they are positioned slightly differently doesn't make a difference to the expected utility calculation, and that is why the states explicitly represented in the matrix justifiably obfuscate this difference. But, of course, the difference between a world where it rains and your friend goes to the movies and a world where it rains and your friend goes to the park makes all the difference to the expected utility calculation. So, why are *these* differences not represented in the matrix?

The simple answer (too simple to be correct, as we shall soon see) is that the subject whose decision we are modeling believes that these worlds will not obtain.[13] The decision a subject faces is determined in part by which states the subject believes

[11] The questions are actually equivalent, for differently grained states are different states. I treat the questions separately anyway because I want to highlight two different sorts of reasons for not including a state in the matrix.

[12] Big differences, as in sandstorms that reach you, can indeed make a difference.

[13] For an argument along these lines, see Weatherson (2012).

have a chance of obtaining. Of course, the subject may in fact believe that there is some chance of his friend's not going to the movies even though it rains—but, in that case, the matrix presented above just doesn't represent the choice that this subject faces. And no subject will think that every possible world has a chance of obtaining. For one, decisions made by subjects who did think that would be incredibly complicated—and they often are not. And it's not just that as a matter of fact subjects make decisions that are not incredibly complicated, but often there is nothing wrong with them doing this, and our decision theory should recognize it. This ruling out of some possible worlds as having no chance of obtaining is not just made for the sake of making our choices easier. I really am justified in ruling out possible worlds where if I eat a slug right now then a genie will give me untold riches. This is not to say, of course, that I think those worlds are metaphysically impossible. Not all possibilities are actual, of course, and these possible worlds are among the possibilities that I (rationally) believe will not obtain.

An alternative is to think that it is never rational to completely rule out anything we take to be a metaphysical possibility. This amounts to the converse of the axiom of normality, and it goes by the name of "regularity." According to normality, logical truths receive probability 1 (and so their negations receive probability 0). According to one version of regularity, one should assign credence 1 *only* to logical truths (and so no logical possibility should be assigned credence 0). However, this conflicts with conditionalizing on (non-logically trivial) evidence. On a more plausible version, only ur-priors are subject to that constraint. Hájek (ms) thinks that the most plausible version of the regularity constraint is one according to which we should assign positive credence to anything that is compatible with what we believe (or should believe). This version of regularity, however, is not in conflict with the view just proposed: it just *is* that view. Hájek ends up nevertheless rejecting even this version of regularity. Since Hájek's preferred version of regularity just is the view presented in the previous paragraph, we better see why he rejects it.

To present Hájek's arguments we need to go back to thinking of probability in set-theoretic terms, as we did in chapter 2. More precisely, mathematicians think of probability as a *triple* $<\Omega, F, P>$. In that triple, P is the familiar probability function, and Ω is a set of events—P will be defined over subsets of Ω. However, P need not be defined over the set of *all* subsets of Ω, called the powerset of Ω, and symbolized as follows: $P(\Omega)$. Rather, the job of F is to pick out a privileged subset of $P(\Omega)$, and P is defined over F. Hájek generalizes the notion of regularity to a kind of congruity between all three items in the triple. One kind of regularity obtains if $F = P(\Omega)$, for then F recognizes everything that is a possibility according to Ω. A second kind of regularity obtains if $P(S) > 0$ for every singleton S member of F.[14] This means that P blesses with positive probability every

[14] Given that the probability P assigns to non-singleton members of F is determined by the probability it assigns to the singletons, this means that P assigns positive probability to every member of F.

possibility recognized by F. Finally, if these two kinds of regularity obtain, then a third one will also automatically obtain: P will bless with positive probability every possibility recognized by Ω.

Hájek's misgivings about regularity parallel his misgivings about what he calls "the ratio analysis" of conditional probability, rehearsed in chapter 2. Assume that we are, once again, throwing darts with point-sized tips at the $[0, 1]$ interval. Certain possibilities, such as the possibility that a dart spontaneously combusts, or quantum-tunnels to China, etc., are not even countenanced in Ω. This might, however, be dismissed as a mere simplifying assumption made for the sake of the modeler's convenience. There are other real possibilities, however, which cannot be so dismissed. There will be subsets of $[0, 1]$ that are *non-measurable*: the assumption that they receive *any* probability leads to a contradiction. Vitali sets are examples of such non-measurable sets.[15] The raison d'être of F is precisely to filter out such subsets from Ω so that they are not even considered by P, and so F is the largest subset of $P(\Omega)$ all of whose members are measurable. Take, then, the possibility that the dart falls on one of the points in a Vitali set. That is a possibility that is definitely countenanced by Ω, but not by F.

Moreover, there are some possibilities that do make it into F but that get assigned a probability 0 by P. All but at most a finite subset of the singletons in F need to get assigned zero probability, and so the second kind of regularity which we introduced in the previous paragraph is guaranteed not to obtain. This is because assigning positive probability to even a countable infinity of singletons leads, by countable additivity, to assigning probability greater than 1 to the whole interval. For the same reason, we cannot assign positive probability to the finite subsets of F—there are uncountably many of them, and so again we would have to assign probability greater than 1 to the whole interval. And the same goes for all the countable subsets of F—there are uncountably many of them too. Indeed, there are even some uncountable subsets of F that must also receive probability 0—for instance, Cantor's "ternary set."[16]

When Ω is uncountably large, then, violations of regularity are inevitable. What does this mean for the view that the states depicted in a (well-constructed) decision matrix are precisely those that are compatible with what the subject believes? It means, simply put, that the view is wrong. Not all the states that are not accounted for on a properly constructed matrix for a decision problem will be incompatible with what the agent believes. Some of them will not appear in the matrix because they correspond to unmeasurable subsets of Ω. We can discard these, if we want, by insisting that the states be taken from F and not directly

[15] See chapter 2 for a brief explanation of Vitali sets.

[16] Cantor's ternary set is obtained by iteratively deleting the open "middle third" from a line segment. Thus, we first delete $(1/3, 2/3)$ from $[0, 1]$, leaving us with the set of points $[0, 1/3] \cup [2/3, 1]$. Then we delete the middle thirds of each one of those two lines, etc. The result is a set with uncountably many members but measure 0.

from Ω. But even some states accounted for in F will not appear in the matrix even if they are compatible with what the agent does (and even should) believe. These will be states that need to be assigned probability 0 for formal reasons, and not because the agent discards them.

But remember why we made the claim that the states that are not represented in the matrix are precisely those incompatible with what the agent believes: we were looking for a place for beliefs in the Bayesian framework. And even if beliefs are not the *only* factors that can determine that a state is not represented in the matrix, they can still be *one* such factor. Sometimes a state will not be in the matrix because it represents a state that, for formal reasons, must be assigned credence 0 (or cannot be assigned any credence at all). But that is not what happens, for instance, with the state where it rains and your friend goes to the park. There is no formal reason to exclude that state from the matrix. Rather, it is excluded because it is incompatible with what you believe. Notice, in particular, that when it makes sense to model the underlying space finitely, then there will be no formal reasons for assigning zero probability to any of the states, and so we can be sure that, in such cases, the reason why any state receives credence 0 in a properly constructed decision matrix is that it is incompatible with the subject's beliefs.

There is a place for beliefs, then, even in the graded framework of Bayesian decision theory. But while what a subject believes determines which decision problem he *in fact* faces, according to Fumerton's thesis it doesn't determine which decision problem he *should* face. Which decision problem a subject *should* face is determined not by what he believes, but by what he is justified in believing. Now, there is an ambiguity in the claim that the right shape of a decision problem is determined by what the subject is justified in believing—it's the ambiguity between the *ex-ante* and the *ex-post* sense of justification explained before. Which propositions are the ones that determine the decision problem that the subject should face, the ones he has *ex-ante* justification for believing or those he is *ex-post* justified in believing? It may happen that a subject has only justified beliefs, and that these beliefs determine a certain decision problem he faces, but he doesn't have *all* the beliefs he is justified in having—and those can determine a *different* decision problem.[17] Consideration of the trolley case presented above suggests that although there is no action that it is *ex-post* rational for the subject to perform in the absence of relevant justified attitudes (and, in this sense, there is no problem he should face), there is still an *ex-ante* rational thing to do, namely, to not pull the switch (and, in this sense, the problem he should face is the one determined by the beliefs he should have).

I have argued, then, that unjustified attitudes do not justify, and that there is a place for justified beliefs in traditional decision theory. The kind of

[17] Arguably no subject believes everything he has *ex-ante* justification for believing, but in most cases this will not affect which decision problem he faces.

knowledge-based decision theory to be presented in the next section does not disagree with this claim—indeed, it entails it, if knowledge entails justification. But knowledge-based decision theory does reject the second component of the view announced at the outset: that rational action tolerates false belief. Therefore, if rational action does tolerate false belief then knowledge-based decision theory must be wrong. I argue that indeed it is in sections 4.5 and 4.6.

4.4 Knowledge-Based Decision Theory

Recall that the kind of knowledge-first Objective Bayesianism presented at the end of chapter 3 posits a unique ur-prior Cr_u, identifies the subject's evidence at a time with what the subject knows at that time, and identifies the justified credences of a subject at a time with the conditional credences given the evidence, as provided by Cr_u, $Cr_u(-|E)$. We can combine this knowledge-first Objective Bayesianism with traditional decision theory. In section 4.3 I argued for the view that whereas the agent's actual beliefs represent the actual decision problem that he is facing, they do not necessarily represent the one he would face were he to be rational. The action that maximizes expected utility relative to the agent's actual beliefs and credences need not be the action it is rational for the agent to perform, if some of those beliefs or credences are not themselves rational. Knowledge-based decision theory has it that unjustified attitudes do not rationalize action, but what it requires in addition is not just rational belief, but knowledge. Thus, according to knowledge-based decision theory, the states that should be represented in a decision matrix are those that are compatible with what the subject knows.

Now, as I already mentioned (in footnote 2), Williamson himself seems to think that the identification of rationality with knowledge follows from the identification of evidence with knowledge. It is obvious that there is no simple logical connection here: one may buy Factualism and also buy the evidentialist claim that all justified beliefs are justified by one's evidence[18] and yet resist the claim that only known beliefs are justified. Consider, for instance, any case of inductive justification. Suppose that I know the premises of a good inductive argument, and on that basis come to believe its conclusion. By calling the inductive argument "good" I mean to imply that knowledge of the premises can give one justification for believing the conclusion. But, of course, knowledge of the premises of an inductive argument is compatible with the falsity of the conclusion. Therefore, it can be rational for me to believe a false proposition, from which it follows that we cannot identify rationality with knowledge. Williamson is free, of course, to reject the characterization of the goodness of an inductive argument that I just gave,

[18] I will reject this evidentialist view in chapter 9.

holding that an inductive argument can be good in the sense of justifying belief in its conclusion only if that conclusion would thereby be known (and, therefore, only if that conclusion is true). But then the identification of rationality with knowledge would follow not just from Factualism, but in addition from this additional thesis (which I discuss further in section 4.5).

There is, however, a route from Factualism, or E = K, to R = K—although it's not one that Williamson himself or any other defender of knowledge-first epistemology has, to my knowledge, put forward. The route gives an answer to the question of which states should be considered in a decision matrix. The answer is different from (though not incompatible with) the view that only states compatible with what the subject is justified in believing should be represented. The alternative answer is that only states compatible with the subject's evidence should be represented. These two answers, as I said, are not incompatible, and indeed taken together with Factualism they imply E = R. If the states that should be represented are those compatible with R, and also those compatible with E, then E = R. Given E = K, then, R = K.

Knowledge-based decision theory is, then, precisely that combination of views: Factualism together with the claim that the states that should be represented in a decision matrix are those that are compatible with your evidence. In the next two sections I take for granted the second component of the view—that your evidence determines which states you should consider—and argue against the first one.

4.5 Not All Knowledge Is Evidence

Some of our knowledge is *inferential*—it is acquired by inference from some other bits of knowledge. And, at least on the face of it, some of that inferential knowledge is *ampliative*: the other known propositions it is inferred from do not entail the inferred proposition. If this is true, then not all knowledge can be evidence, because not all inferred knowledge can be evidence. Now, this problem for Factualism interacts in interesting ways with a fundamental problem in epistemology—a version of what Stewart Cohen called "the problem of easy knowledge." My preferred solution to the problem of easy knowledge conflicts with the claim that some inferential knowledge is ampliative (at least on some understandings of what it takes for an inference to be ampliative). So it might look as if there is really no problem for Factualism, for one of the premises on which the objection rests is false (or, at least, controversial). In this section I explain the problem for Factualism assuming that there is ampliative knowledge. I take up that assumption in chapter 8, where I deal with the easy knowledge problem and show why even lifting the assumption that there is knowledge by ampliative inference doesn't save Factualism from the problem presented in this section.

Let us start by assuming, then, that there can be ampliative inferential knowledge. Abstractly, let us suppose that a subject can come to know that Q on the basis of his knowledge that P even though P does not imply Q. A crucial question regarding this case is: what credence should the subject have in Q? Assuming that P is the *only* evidence the subject has which bears on Q, the answer that Objective Bayesianism gives, of course, is that it should be $Cr_u(Q|P)$. Now, $Cr_u(Q|P) \leq Cr_u(P \rightarrow Q)$, that is to say, conditional credences are bounded above by credences in the corresponding material conditionals.[19] Given that P doesn't entail Q, $P \wedge \neg Q$ is a logical possibility. Thus, there is at least pressure to say that the subject shouldn't rule it out (although I will examine this assumption more closely momentarily). If so, then $Cr_u(P \wedge \neg Q) < 1$, and so $Cr_u(Q|P) < 1$, although presumably they will be close to 1. Consider now this question: what should the subject's credence in Q be after he learns that P and, on its basis, comes to know that Q? There are two relevant conditional credences here. One is the credence in Q conditional on P—the proposition on the basis of which the subject knows that Q. The other is credence in Q conditional on everything the subject knows at the relevant time (after he makes the inference to Q). Now, if Factualism is true, then part of the subject's evidence at the relevant time is Q itself. And the subject's credence in Q, according to Factualism, should be 1.

But a credence of 1 in Q is irrational overconfidence. One's credence in a proposition that one arrived at inferentially should depend only on the evidence from which it is inferred and the quality of the inference. Suppose that you know that there are two balls in an urn, and that each one is either black or white. Now suppose that you take one out, notice its color, and put it back in the urn. You repeat this procedure ten times, and every time the ball is white. I would take myself to know, on this basis, that both balls are white. However, I should not assign credence 1 to the proposition that both balls are white.

This is the bedrock problem for Factualism, but it can be dramatized by appealing to the fact that this irrational overconfidence can have bad downstream consequences. There are bound to be propositions Q and R such that $Cr_u(R|P) \neq Cr_u(R|Q)$, and Factualism wrongly entails that the rational credence in R, after one comes to know that Q only by inferring it from P, is given by the latter. For instance, I shouldn't be certain that the white wall in front of

[19] The following proof is due to Jason Turner (see Calabrese (1987) for an alternative proof): by the ratio formula, $Cr_u(\neg Q|P) = \frac{Cr_u(\neg Q \wedge P)}{Cr_u(P)}$. Notice that $Cr_u(\neg Q \wedge P) \geq Cr_u(\neg Q|P)$, for if both m and n are positive real numbers less than or equal to 1, then $\frac{m}{n} \geq m$. Multiplying both sides of that inequality by -1 changes its direction: $-Cr_u(\neg Q|P) \leq -Cr_u(\neg Q \wedge P)$. Adding 1 to both sides gives us $1 - Cr_u(\neg Q|P) \leq 1 - Cr_u(\neg Q \wedge P)$. An elementary theorem of probability theory is that the probability of the negation of a proposition is 1 minus the probability of that proposition. So $1 - Cr_u(\neg Q|P) = Cr_u(Q|P)$ (for conditional probabilities are probabilities) and $1 - Cr_u(\neg Q \wedge P) = Cr_u(\neg(\neg Q \wedge P)) = Cr_u(\neg P \vee Q)$. Therefore, $Cr_u(Q|P) \leq Cr_u(\neg P \vee Q)$, and so $Cr_u(Q|P) \leq Cr_u(P \rightarrow Q)$.

me is the same color as either of the balls in the urn. Moreover, these irrational credences can have downstream effects not only on further credences, but also on actions. Suppose that something important turns on whether the two balls are the same color. In that case too, Factualism has the wrong consequence that the subject should perform the action rationalized by his irrational overconfidence.[20]

Williamson himself appeals to these kinds of downstream effects when criticizing alternatives to Factualism:

> If evidence required only justified true belief, or some other good cognitive status short of knowledge, then a critical mass of evidence could set off a kind of chain reaction. Our known evidence justifies belief in various true hypotheses; they would count as evidence too, so this larger evidence set would justify belief in still more true hypotheses, which would in turn count as further evidence. The result would be very different from our present conception of evidence.
>
> (Williamson (2000), p. 201)

But this passage is puzzling because the criticism Williamson presents here applies equally well to Factualism. As Goldman put it:

> I am puzzled that Williamson presents this argument against the JTB [justified true belief] account of evidence, because the contemplated chain reaction seems almost as threatening under the E = K construal. Clearly, E = K does not restrict evidence to non-inferential knowledge. So if one's 'basic' evidence justifies belief in various true hypotheses that are also known, a very similar chain reaction looms.
>
> (Goldman (2009), p. 88)

The objection to Factualism that I just presented relies on the claim that $Cr_u(Q|P) < 1$, and that claim relied in turn on the observation that the subject should not rule out the possibility that $P \wedge \neg Q$—and the fact that $Cr_u(Q|P) \leq Cr_u(P \rightarrow Q)$. But why shouldn't the subject rule out that possibility? It is a logical possibility, to be sure, but there are two obstacles to the claim that Cr_u shouldn't rule out any logical possibilities: the anti-regularity arguments of Hájek presented in section 4.3, and the problem of logical omniscience discussed in chapter 2. I take these in turn.

[20] These consequences may be avoided by appealing to something like the knowledge-to-action principle defended by Fantl and McGrath—see Fantl and McGrath (2002), Fantl and McGrath (2007), and Fantl and McGrath (2009), and see Comesaña (2018a) for an explanation of how those principles can help here. Williamson himself does not like principles like those, and appeals instead to the alleged failure of knowledge to iterate to handle the kind of cases we are interested in (see Williamson (2005)). On whether knowledge iterates, see Cohen and Comesaña (2013a) and Cohen and Comesaña (2013b), as well as Greco (2015a) and Greco (2015b).

The reasons Hájek has for being weary of regularity are all, recall, formal reasons. They arise in the context of a probability measure defined over an uncountable set. Some subsets of these sets will not make it into the probability field *F*, for they cannot coherently be assigned *any* probability, and some other subsets will have to be assigned probability 0, on pain of violations of (countable) additivity. But the propositions in question in cases of inferential knowledge are unlikely to be those kinds of propositions. And if they are, then although this may "solve" the problem for knowledge-based decision theory, it would do so only by creating others. For now we can no longer assume that the states that should be considered in a decision problem are all and only those compatible with the subject's evidence—there will be certain states that have to be ruled out even though they are compatible with the subject's evidence. The problems with regularity, then, are no help to Factualism.

In our discussion of the problem of logical omniscience in chapter 2 I argued that logical omniscience is not a rational requirement, and thus that we should be careful when applying the Bayesian framework not to derive consequences from it which depend on this aspect of the models. But am I not assuming logical omniscience when arguing that *$Cr_u(P \wedge \neg Q) > 0$ because it is a logical possibility*?

I am not assuming logical omniscience. I am only assuming that, in at least some cases where the subject gains knowledge by means of an ampliative inference, the subject knows that the inference in question is ampliative. Notice what the friend of Factualism will have to claim in order to get out of the problem: that, in every case in which a subject gains knowledge by means of an ampliative inference, the subject thinks that the inference is not ampliative. This is extremely implausible.

I conclude, then, that inferentially acquired knowledge is not evidence, and, thus, that not all knowledge is evidence.

4.6 Not All Evidence Is Knowledge

The problem for Factualism presented in the previous section can be solved by replacing E = K with E = basic K: only knowledge is evidence, but only basic (non-inferential) knowledge is evidence.[21] The problem I present in this section challenges the other direction of the equation: I argue that not all evidence is knowledge—and, in particular, that some false propositions are evidence.

My argument takes a familiar route: a consideration of two cases, a good case where the subject has knowledge and a corresponding bad case where the

[21] For a similar suggestion, see the last paragraph of Bacon (2014).

subject lacks knowledge but everything is otherwise as similar as possible to the good case:

Good Lucas: Tomás would like to eat some candy. Lucas offers him some, and Tomás reaches for it and puts it in his mouth. Everything goes as planned and Tomás enjoys some candy.

Bad Lucas: Tomás would like to eat some candy. Lucas offers him a marble that looks just like candy, and Tomás reaches for it and puts it in his mouth. Tomás is disappointed.

Let us first look at this pair of cases through the eyes of a knowledge-based decision theorist. According to that view, the states that must be represented in the matrix for a decision problem are all and only those states compatible with what the subject knows. In section 4.5 I argued that propositions that the subject knows through ampliative inference shouldn't be allowed to determine the relevant states. But to bypass this issue, let us suppose that when Tomás knows that Lucas is offering him candy he knows it non-inferentially, or at least not through an ampliative inference. Then, knowledge-based decision theory has the consequence that the state where Lucas is offering a marble to Tomás should be part of the decision matrix for Bad Lucas but not for Good Lucas. This means, in turn, that according to knowledge-based decision theory Tomás's action of tasting what Lucas gave him is rational in Good Lucas but irrational in Bad Lucas—tasting a marble, which is the outcome of tasting what Lucas gave Tomás in Bad Lucas, has very low utility for Tomás.

This is the wrong result. Tomás's action of reaching for the thing Lucas is offering him and tasting it is equally rational in both cases. After all, the rationality of an action cannot depend on features of the situation that are inaccessible to the subject himself. It is true that Tomás would have been better off had he not tasted the marble in Bad Lucas, but such is the fate of us fallible creatures: sometimes the rational action is not the one that results in the best outcome. Everyone, knowledge-based decision theorists included, accepts this distinction between rationality and optimality in at least some cases: actions that maximize expected utility don't always maximize utility.[22] Professional poker players are well aware of this distinction, and even have a name for the tendency to confuse rational and

[22] Moreover, even knowledge-based decision theorists should accept that acting rationally may sometimes entail performing an action *you know* to be suboptimal, as witnessed by the mine shafts case by Reagan (1980), discussed in Parfit (2011). To use Jacob Ross's version of the case in Ross (2012), suppose that you are presented with three envelopes: A, B, and C. You know that envelope A contains $900, and that one of the other two envelopes contains $1,000, and the other one is empty (you don't know which is which). You can choose one envelope and keep its contents. It is, of course, rational to choose A, even though you know that one of the other envelopes is guaranteed to have more money in it.

successful action: "resulting."[23] And yet, isn't the knowledge-based decision theorist resulting when he says that Tomás's action was irrational?

Before developing my argument in some more detail, I want to briefly comment on a similar argument from Julien Dutant (forthcoming). Dutant argues against knowledge-based decision theory on the basis of pairs of cases like the one above (indeed, the name "knowledge-based decision theory" is his). In place of knowledge-based decision theory, Dutant proposes a two-step account according to which what the subject knows fixes which propositions are supported for the subject, and the subject's evidence at a time consists of the propositions that are supported for the subject at that time.[24] Although I agree with Dutant's criticism of knowledge-based decision theory, I do not think that his two-step account fares any better, for two reasons. First, Dutant is in effect proposing that the subject's evidence contains some propositions that are inferentially acquired. This amounts to saying that it is permissible (indeed, obligatory) for a subject to rule out a state only because it is incompatible with a proposition one acquired inferentially. I argued in section 4.5 that this is implausible in the case where the inferentially acquired belief amounts to knowledge, and it is all the more implausible in the cases to which Dutant applies his view, where the inferentially acquired belief does not amount to knowledge. Second, Dutant's two-step account is vulnerable to the same kind of counterexample as knowledge-based decision theory is in the special case where the proposition in question is known non-inferentially if at all. Indeed, given our assumption that Tomás knows that Lucas is offering candy non-inferentially if at all, the pair of cases presented above are a counterexample to Dutant's account as much as to knowledge-based decision theory.

Dutant could point to knowledge Tomás has which supports the propositions that Lucas is offering candy even in Bad Lucas (such as, say, the proposition that Lucas is offering something that looks like candy). But notice that if Tomás knows that what Lucas is offering looks like candy, he also knows that what Lucas is offering looks like a marble that looks like candy. Moreover, Tomás's belief is not based on these pieces of knowledge. Why would it be rational for him to rule out the possibility that Lucas is offering a marble, if that ruling out is not performed on the basis of anything he knows? The natural answer is that the ruling out is nevertheless justified, but that is not an answer that Dutant can give.

Let us now return to Bad Lucas. I take it as a datum that Tomás acts rationally in that case. My appeal to the inaccessibility of the fact that Tomás doesn't know that Lucas is offering a marble is more of an explanation of that datum than an argument for it. However, claims of accessibility are at least somewhat metaphorical, and so we better be clear on what the claim amounts to.

[23] See Duke (2018).
[24] For similar views, see Lord (2018) and Kiesewetter (2017).

I start with a clarification as to what the accessibility claim does *not* amount to: it does not amount to an infallibility claim. I am not saying that the rationality of an action can only depend on features to which we have infallible access. Williamson has rightly pointed out that just because we can make mistakes in following a rule that doesn't mean that we are following a different rule when we don't make a mistake (Williamson (2000), pp. 191–2). So the claim is not that because Tomás mistakenly believes, in Bad Lucas, that Lucas is offering him some candy, then his reason for acting in Good Lucas cannot be that Lucas is offering him some candy.[25]

A first pass at a positive clarification of the accessibility claim is the following. The difference between Good Lucas and Bad Lucas consists merely in the truth-value of the proposition that Lucas is offering some candy to Tomás—and everything that supervenes on that difference.[26] But the mere truth of a proposition like that one cannot make a difference as to which action is rational.[27]

Williamson will disagree. He will rightly point out that another difference between Bad Lucas and Good Lucas is that in the former case, but not in the latter, Tomás knows that Lucas is offering candy. So the difference is not just in the truth-value of the proposition regarding Lucas's offering, but also in Tomás's knowledge. And whereas it might be right that a mere difference in the truth of the proposition that Lucas is offering some candy cannot make a difference as to which action is rational for Tomás to perform, which propositions Tomás knows is intuitively very relevant to which action is rational for Tomás to perform. I said above that the only differences between Good Lucas and Bad Lucas are in the proposition regarding Lucas's offering—*and in everything that supervenes on that.* Given that whether Tomás knows that Lucas is offering candy is one of the differences that supervenes on whether Lucas is in fact offering candy (for we are holding fixed other relevant factors), the difference in the truth-value of that proposition does in this case make a difference as to which action it is rational for Tomás to perform—or so the knowledge-firster will argue.

But although intuitively what the subject knows can affect which actions are rational for him to perform, pairs of cases like the one we are considering show that this is not always the case. Sometimes, whether the subject knows a proposition depends exclusively on whether that proposition is true. In those cases,

[25] Indeed, I am so far from making this claim that, in my own view, Tomás's reason for acting, both in Good Lucas and in Bad Lucas, is that Lucas is offering him some candy (even though that reason is false in Bad Lucas). See chapter 6.

[26] It's important to note here that I am not conceiving of the relationship between Good Lucas and Bad Lucas as a counterfactual-like relation, where, for instance, Good Lucas is the closest case to Bad Lucas where Tomás knows. The crucial point is that, however we conceive of these cases, the subject in the bad one cannot be justified in believing that the cases are different.

[27] I say that the mere truth of a proposition "like that one" cannot make a difference as to which action is rational in anticipation of the reply that, of course, there are propositions such that a mere change in their truth-value can affect which action is rational, such as the proposition "Action A is rational."

whether the subject knows the proposition cannot make a difference to which actions are rational for him. For the interesting change—that the subject ceases to know a certain proposition—is due exclusively to a boring change—that the proposition ceases to be true.[28] Paraphrasing Stich (1978), what knowledge adds to rational belief is not always epistemically relevant.

That the features of the situation that make an attitude rational must be accessible to the agent in question is related to another, also somewhat tired and overused metaphor: that norms of rationality must be action-guiding. I propose the following as a first-pass, minimalist interpretation of that metaphor: if a condition *C* is such that its obtaining makes some attitude or action on the part of a subject irrational, then it must be the case that it is rational for the subject to believe that *C* obtains. Notice that the assumption does not say that the subject must actually believe that *C* obtains, only that it is rational for the subject to believe that it obtains. But the condition is not the very weak one that there must be a possible circumstance wherein it is rational for the subject to believe that *C* obtains—this would make it an almost trivial constraint, because, for almost any condition, there is some circumstance or other such that it is rational to believe, in that circumstance, that the condition obtains. Rather, what must happen is that it is rational for the subject to believe that *C* obtains when the fact that *C* obtains is what makes the attitude or action irrational. So, applied to our case, which action it is rational for Tomás to perform in Bad Lucas cannot depend on the fact that he doesn't know that Lucas is offering candy, for Tomás cannot rationally believe that he doesn't know that Lucas is offering him candy.[29]

It is well worth repeating that I am not saying that it is impossible for Tomás to rationally believe that he doesn't know that Lucas is offering him candy. If the marble didn't look very much like a piece of candy, or if Lucas had a track record of playing these kinds of tricks on Tomás, then it might well be rational for Tomás to believe that he doesn't know that Lucas is offering him candy. Rather, what I am saying is that it is not rational for Tomás to believe that proposition in the situation as described in Bad Lucas. More generally, when we consider a situation where knowledge is present and then a corresponding situation where it is absent but everything else, as far as the subject knows, is left untouched as far as possible, it will not always be the case that the lack of knowledge in that corresponding situation is rationally believable by the agent. I am also not denying that Tomás's reason for action in Good Lucas is that Lucas is offering him candy. Indeed, as anticipated above, my own view is that Tomás's reason for acting is precisely that proposition in *both* Good Lucas and Bad Lucas.

[28] See Comesaña (2005a) for related discussion.

[29] According to me, of course, it is rational for Tomás to believe that he does know. What is the rational attitude according to the proponent of knowledge-first decision theory? That is a very interesting question, which I address in chapter 5.

Notice, relatedly, that whether a condition is accessible (in my explanation of that notion) can display the kind of asymmetry familiar from consideration of skeptical scenarios. In Good Lucas, Tomás knows that Lucas is not offering him a marble. So, in Good Lucas Tomás knows that he is not in Bad Lucas. So, in Good Lucas, Tomás can discriminate Good Lucas from Bad Lucas. In Bad Lucas, however, Tomás does not know that he is not in Good Lucas (in fact, according to me he rationally believes that he is in Good Lucas). So, in Good Lucas, Tomás cannot discriminate Good Lucas from Bad Lucas. This explains why it is irrational for Tomás to believe, in Bad Lucas, that he doesn't know that Lucas is offering him candy.

So, the first pass at my interpretation of the metaphors of accessibility and guidance was the following: if a condition *C* makes it the case that a subject is irrational, then it must be rational for that subject to believe that *C* obtains. Why did I call it a first pass? Because it might need some amount of Chisholming in order to make it counterexample-free. For instance, suppose that a subject believes a proposition on the basis of evidence which doesn't justify belief. Then, arguably, the fact that the evidence on which he bases his belief doesn't justify that attitude makes it the case that his belief is unjustified. Must it be the case that the subject is justified in believing that the evidence makes his belief unjustified? Some authors think that the answer is "Yes." For instance, Titelbaum (2015) argues for the thesis that false beliefs about the rules of rationality are themselves irrational. If Titelbaum is correct, then if the evidence does not support belief it is rational to believe that it doesn't support belief. Similarly, Smithies (2011) argues for very strong theses which imply that if it is irrational to believe *p* then it is rational to believe that it is irrational to believe *p*. I am not wholly unsympathetic to strong accessibility theses such as Titelbaum's and Smithies's,[30] but I do not think that the plausibility of my argument against knowledge-based decision theory depends on them. I would therefore like to reformulate my interpretation of the accessibility and guiding metaphors so as to avoid commitment to such theses. One way to do it would be to restrict the principle to cases where *C* is, in some appropriate sense, non-epistemic. This would rule out cases such as the one just canvassed. Of course, it would then behoove me to clarify the sense of "epistemic" at issue, a task that I do not welcome. Rather, I rest content with pointing out that merely hurling counterexamples at the principle does not amount to a defense of knowledge-based decision theory from my argument. According to knowledge-firsters, what makes it the case that it is irrational for Tomás to taste what Lucas is offering him in Bad Lucas is that Tomás doesn't know that Lucas is offering a candy, and Tomás doesn't know that Lucas is offering a candy because Lucas is offering a marble. Without assuming the general transitivity of the *making-it-the-case*

[30] Indeed, I argued for at least the interest of models of epistemic logic that validate S4—see Cohen and Comesaña (2013a).

relation, it seems plausible that this means that what makes it the case that it is irrational for Tomás to taste what Lucas is offering is that Lucas is offering him a marble. My objection is that it is irrational for Tomás to believe that Lucas is offering him a marble, and it is conditions *like that one* that I mean to be possible values for *C*. It may well prove difficult to specify the relevant class here, but that doesn't mean that we cannot recognize members of the class when we see them.

Notice that standard cases of defeaters fit the mold of my principle. Suppose that I have an experience as of a red table in front of me, and on this basis I believe that there is a red table in front of me. The proposition that there are red lights shining on the table can make it the case that my belief is irrational only if it is rational for me to believe it. Even if in fact there are red lights shining on the table, if it is not rational for me to believe that there are, that fact does not make my belief that the table is red irrational.[31]

It is important that my cashing out of the metaphors of accessibility and guiding is done in terms of rational believability, not in terms of knowability. Williamson would reject an interpretation in terms of knowability. He thinks that we are not always in a position to know what our evidence is, but that this is not a problem for E = K, given that there is no (non-trivial) condition that is "luminous" in the sense that, whenever it obtains, we are in a position to know that it obtains (see Williamson (2000), chapter 4). Williamson's anti-luminosity argument makes crucial appeal to the notion of safety, however, enshrined in the principle that if one knows that p then one truly believes that p in all nearby circumstances.[32] Such a principle, however, will be plainly unacceptable if we replace *knowledge* by *rational belief.* If it is possible to have false rational beliefs, it is obviously possible to have unsafe rational beliefs. Of course, a proponent of R = K will reject the claim that it is possible to have false rational beliefs, but an appeal to that view in this context will be plainly question-begging, if the argument for R = K is to go through E = K.

4.7 Conclusion

In this chapter I have argued against knowledge-based decision theory and, via the connection between decision theory and epistemology afforded by Fumerton's thesis, against Factualism. Traditional (Bayesian) decision theory has it that the states that a subject should consider in making a decision are all of those

[31] For more on defeaters, see chapter 6.

[32] I have argued (see Comesaña (2005b)) that a subjunctive analysis of the "nearby" clause in the definition of safety does not give us a necessary condition on knowledge. Williamson himself does not go for such a subjunctive analysis, or for any other analysis for that matter, for he is content to let our judgments about safety to go hand in hand with our judgments about knowledge. See also Cohen and Comesaña (2013a) for a defense of KK in light of Williamson's argument.

compatible with his evidence. Fumerton's thesis, on one formulation, is the claim that the states that a subject should consider in making a decision are those that are incompatible with what the subject is justified in believing—and so, that the subject's evidence consists of all those propositions he is justified in believing. Fumerton's thesis must face objections from two sides. On the one hand, Subjective Bayesianism, although with no official account of evidence to offer, has a natural companion: the idea that a subject's evidence at a time is provided by what the subject believes at that time. I offered an argument for this position: that, unless it is true, practical dilemmas ensue. In response, I granted that dilemmas *of a kind* ensue, but also argued that the kind in question is benign. On the other hand, knowledge-based decision theory has it that the states that a subject should consider when making a decision are those that are incompatible with what he knows. I presented two main arguments against knowledge-based decision theory. The first one concentrates on the right-to-left direction of E = K. Because some knowledge is acquired by ampliative inference, not all knowledge can be evidence. The second one concentrates on the left-to-right direction. Because some actions are rational even when based on false beliefs, not all evidence can be knowledge. One defense of E = K from this second objection appeals to the distinction between justification and excuses. I take up and argue against that defense in chapter 5.

5
Excuses, Would-Be Knowledge, and Rationality-Based Decision Theory

5.1 Introduction

The structure of one of my arguments against knowledge-first decision theory in chapter 4 is a counterexample: that version of decision theory has the wrong consequence regarding Bad Lucas. I also suggested that knowledge-based decision theory might fail in this way because it doesn't satisfy plausible interpretations of the metaphor that conditions on rational belief should be accessible and action-guiding. I am more confident, however, that knowledge-based decision theory gets some cases wrong than of this diagnosis of why it does so. The central objection to knowledge-based decision theory has repercussions for knowledge-first epistemology more generally. For, against the background of Fumerton's thesis—that an action can be rational only if the attitudes, including the beliefs, on which it is based are themselves rational—the judgment that Tomás's action is rational in Bad Lucas entails that his belief that what Lucas is offering is candy is also rational. But, of course, this belief is false. Therefore, there can be false rational beliefs, and so K = R is false.

Now, some may think that K = R is in any case an overreach on the part of the program of knowledge-first epistemology. My own view is that the problems for K = R presented in chapter 4 motivate the return to a more traditional epistemology and decision theory, ones in which a notion of rationality not explained in terms of knowledge plays a central role. I will present such a view towards the end of this chapter, and develop some of its contours in later ones. But such a path is of course closed to any proponent of knowledge-first epistemology. Many of them, however, are not happy with simply biting the bullet represented by the kinds of counterexamples presented in chapter 4, thinking instead that some degree of damage control is in order.

There are two camps when it comes to how such damage control should proceed. One camp offers an error theory regarding claims like the ones that form the core of chapter 4—that Tomás acts and believes rationally in Bad Lucas. Tomás's actions and beliefs in Bad Lucas are *excusable*, in the sense that Tomás is blameless for so acting and believing, but they are not justified or rational in any epistemically substantive sense of those words. The other camp is more conciliatory, arguing that Tomás's actions and beliefs are indeed rational, but that the

Being Rational and Being Right. Juan Comesaña, Oxford University Press (2020). © Juan Comesaña.
DOI: 10.1093/oso/9780198847717.001.0001

sense of rationality in question can be explained in terms of knowledge. In this chapter I examine both the excuses maneuver as well as the knowledge-first accounts of rationality, and find them both wanting. This further solidifies the return to a rationality-first account of practical and epistemic rationality.

5.2 Excuses and Dilemmas

In this section I argue against the excuses maneuver. I start by pointing out a defect that such maneuvers have in general, and then I analyze in some more detail two specific implementations of it: one by Williamson and one by María Lasonen-Aarnio.

5.2.1 Excuses in General

Any view according to which false beliefs can never be justified must face at least two problems. First, there is what I shall call the demarcation problem. Not all false beliefs are on a par, and even someone who claims that they are all unjustified must account for differences within this class. For instance, Tomás's belief in Bad Lucas is false, but there is a big epistemic difference between it and a paradigmatic case of irrational belief, such as a belief caused by wishful thinking. The problem of demarcation, then, is precisely the problem of how to distinguish the good from the bad cases of false beliefs.

A second problem for any view that denies that there can be false justified beliefs is to explain which attitudes *are* justified towards the appropriate class of false beliefs. For instance, which attitude is Tomás justified in taking towards the proposition that Lucas is offering candy? For reasons that will become apparent shortly, I call this the suspension problem. Notice that a solution to the demarcation problem is not automatically a solution to the suspension problem. A solution to the demarcation problem *may* take the form of specifying a doxastic attitude other than belief that fits the false propositions in question, but it need not.

The suspension problem is important in itself, but without an answer to it a practical problem arises as well. Until we are told which attitude Tomás is justified in taking, we do not know which action he is justified in performing. Therefore, a knowledge-first epistemology without an answer to the suspension problem is not fit to be incorporated into a decision theory.

The suspension problem and its consequential practical problem are urgent because it is not obvious that there are *any* answers available to the knowledge-firster. I exemplify with the epistemic question, but similar remarks apply to the practical question. If it is not rational for Tomás to believe that Lucas is offering candy, then which attitude is it rational for Tomás to take towards that proposition? The knowledge-firster cannot say that it is rational for Tomás to *dis*believe that Lucas is offering candy, for of course Tomás doesn't know that Lucas is not offering him candy either. There seem to be three more possible answers available to the knowledge-firster: that Tomás should suspend judgment, that he should adopt some credence (these two are not incompatible with each other), or that there is no doxastic attitude that is rational for Tomás to adopt towards that proposition. I will argue that all three options are unappealing.

Take first the view that Tomás should suspend judgment on the proposition that Lucas is offering candy.[1] Notice, to begin with, how unnatural this view is. Bad Lucas doesn't look at all like a paradigmatic case where suspension of judgment is justified. Even knowledge-firsters are keen to point out that Tomás's epistemic position with respect to the proposition in question is very strong. For instance, there are many propositions that Tomás knows that should be counted as evidence for the proposition that Lucas is offering candy.[2] But perhaps suspension is compatible with strong epistemic position. There is nevertheless a much more important problem for the suggestion that Tomás should suspend judgment. If we hold that Tomás should believe the proposition in question, then we know what consequences that has for the practical case: Tomás shouldn't even consider the case where his belief is false. But if Tomás suspends then, presumably, he should so consider it. In that case, the obvious question is: how should he consider it? What credence should he assign to the proposition in question? Merely saying that he should suspend leaves this important question unanswered.

A knowledge-firster might say (in addition to or instead of saying that suspension of judgment is the justified attitude) that Tomás should adopt some credence < 1 in the proposition that Lucas is offering candy. That option, however, also runs head on into the argument of the previous chapter. If we agree with the knowledge-firster that it is rational for Tomás to believe that Lucas is offering candy in Good Lucas, and thus to rule out of consideration the state where he isn't, then not ruling out that state in Bad Lucas is irrational underconfidence, and goes against the minimalist cashing out of the guiding metaphor I argued for in chapter 4. Of course, the proponent of the excuses maneuver would say that Tomás is excused for believing and acting as he does in Bad Lucas, and that it is this excusability that our intuitions that Tomás is acting and believing justifiedly

[1] There is some textual evidence that this is not an option that Williamson himself will go for in some (admittedly somewhat cryptic) remarks he makes about Pyrrhonism in Williamson (2017b) and Williamson (forthcoming).

[2] I take the case to be one where the belief is not inferentially acquired, but the distinction between inferential and non-inferential justification has to do with the *basis* for the belief, and Williamson is notoriously silent on the basing condition on *ex-post* justification—despite his surprising remark in Williamson (forthcoming) that "[t]he present concern is with doxastic rather than propositional justification."

are tracking. But this answer confuses the demarcation with the suspension problem. We have been told something good about Tomás's actions and beliefs, so that answers (in the proponent's mind) the demarcation problem. But we are now being told that it is rational for Tomás to adopt an attitude other than belief towards the proposition that Lucas is offering candy, and consequently that it is rational for him to refuse the offering. The problem is that this notion of rationality is not recognizable as such, regardless of whether it is paired with a notion of excusability that plays some of its roles.

The remaining option for the knowledge-firster is to say that there is no rational attitude for Tomás to take towards the proposition that Lucas is offering candy. Given this, and given Fumerton's thesis, there is no rational action for Tomás to perform either. This amounts to saying that, in Bad Lucas, Tomás faces both an epistemic and a practical dilemma. Moreover, the dilemmas in question all fall into the *bad* side of the three distinctions made in the previous chapter.

First, the dilemmas are not merely *ex-post*, but also *ex-ante*. On the epistemic side, it's not just that there is no attitude that Tomás can rationally adopt towards the proposition that Lucas is offering candy, but, independently of whether Tomás adopts it or not, there is no rational attitude for him *to* adopt towards that proposition. On the practical side, it's not just that there is no action that Tomás can rationally perform, but, independently of whether Tomás performs it or not, there is no action that is rational in his situation.

Second, the dilemmas are world-imposed, not self-imposed. There is nothing Tomás did, even in the broadest sense of "doing," such that the dilemmas he finds himself in are the consequence of that action.

Third, and relatedly, the dilemmas are non-recoverable. There is nothing Tomás can do such that his doing that will result in his no longer facing the epistemic and practical dilemmas he faces. In particular, there is no justified attitude he can adopt towards the proposition that Lucas is offering candy such that, were he to adopt that attitude, he would no longer face the dilemmas. This is because, as I just said, assuming Factualism there is no justified attitude to take towards that proposition.

The kind of dilemmas posited by the view under consideration, then, are *bad* dilemmas. It is highly implausible to claim that, every time a subject has what I would say is a justified false belief (and what knowledge-firsters would say is a merely excusable belief), he faces epistemic and practical dilemmas of this kind.

To recap, whereas the excuses maneuver may offer the knowledge-firster some degree of damage control in that it affords them *something* good to say about what are intuitively justified beliefs and actions, it does nothing to answer the question that is embarrassing for them: namely, which attitudes and actions are the ones that are justified, if not the intuitive ones?

That is a problem for the excuses maneuver in general. In the rest of this section I examine two particular implementations of the maneuver.

5.2.2 Williamson on Justification and Excuses

I start with Williamson (forthcoming). In that paper, Williamson distinguishes between primary, secondary, and tertiary norms.[3] Let us start by considering just the form of the norms, and bracketing for now their content. A primary norm will have the following form:

$$\text{N: } O(B(p) \rightarrow C)$$

In N, O is some kind of rational obligation operator with wide scope over some kind of conditional. That conditional, in turn, places a necessary condition on believing some proposition p. Such a norm amounts to saying that it is irrational to believe that P while not in C.

According to Williamson, any norm of form (N) gives rise to a secondary norm of the following form:

DN: You ought to have a general disposition to comply with N.

(N) also gives rise to a tertiary norm:

ODN: You ought to do what someone who complies with DN would do in your situation.

Even though Williamson calls DN and ODN norms, it is clear that he thinks that only N has real normative force. For your belief will be justified only to the extent that it complies with N, regardless of whether it also complies with DN and ODN. However, complying with DN and ODN gives you an *excuse* to not comply with N.

The structural point that Williamson wants to make is that we shouldn't confuse satisfaction of a secondary or tertiary norm with satisfaction of a primary norm. Of course we shouldn't. But how is this relevant to whether false beliefs can be justified or merely excusable? This is what Williamson seems to be thinking. According to him, the content of the primary norm N has to do with knowledge:

$$\text{K: } O(B(p) \rightarrow K(p)),$$

which amounts to saying that it is irrational to believe what you don't know. K gives rise to the following secondary and tertiary norms:

[3] In Williamson (2017b), Williamson argues that there are two notions of rationality: a content-based notion and a disposition-based notion. The differences with the paper I discuss in the main text do not seem to me to make a difference to my arguments in this section.

DK: You ought to have a general disposition to comply with K.
ODK: You ought to do what someone who complies with DK would do in your situation.

In Cohen and Comesaña (2013b) Stewart Cohen and I argued, against Williamson, that it can be rational to believe false propositions—for instance, that it can be rational to believe that there is a red table in front of you when there isn't but you are hallucinating it. Williamson (2017b) thinks that views like ours arise out of confusing these kinds of norms.[4] In particular, even though someone who has a false belief may well be complying with DK and ODK, he will necessarily not be complying with K. But one need not confuse norms to hold our view. Indeed, we can formulate the position defended in the paper Williamson is attacking in Williamson's own framework. In that paper we argued for the following as a fundamental norm:[5]

$$\text{E: } O\big(B(p) \rightarrow E(p)\big),$$

where $E(p)$ means that the subject has sufficient evidence for p. We can, of course, recognize at least the conceptual distinction between E and the corresponding secondary and tertiary norms it gives rise to:

DE: You ought to have a general disposition to comply with E.
ODE: You ought to do what someone who complies with DE would do in your situation.

Williamson's argument works only if the correct primary norm is K as opposed to E. In that case, thinking that there can be false rational beliefs *could* be the result of confusing K with DK or ODK. But what is at issue in the dialectical context is precisely whether anything like K could be the primary norm of belief. In the paper Williamson cites, we argued for something like E instead. We do not disagree with Williamson on the distinction between the different *kinds* of norms; we disagree with him on the *content* of those norms. As we say in Cohen and Comesaña (forthcoming), "to disagree with Williamson about how to apply the distinction between justification and excuses is not to fail to take into account the distinction."

[4] Williamson also makes the correct point that distinguishing between the different kinds of norms opens the door for saying that false beliefs never satisfy the primary norm. This is true, but irrelevant with respect to the question of what the content of the primary norm is. In Williamson (2017b), where the distinction between types of norms gives way to a distinction between two kinds of rationality, $rationality_{cont}$ and $rationality_{disp}$, Williamson claims that to hold that there can be false rational beliefs "is to equivocate between what is $\text{rational}_{\text{cont}}$ to believe and what is $\text{rational}_{\text{disp}}$ to believe, with all the consequent danger of smoothing out the fallacies" (p. 267).

[5] The view developed in this book does not abide by that evidentialist norm (see chapter 8), but the dialectical point made in the text is independent of commitment to that particular norm.

One aspect of Williamson's proposal which I haven't discussed is his claim that, in every case (or the most central cases) where we want to say that someone has a false justified belief, the subject in question satisfies a norm of the form DN or ODN, even if he fails to satisfy a norm of the form N. Williamson himself doesn't get too deep into the details of why he claims this, but Maria Lasonen-Aarnio does. I turn now to her deployment of the excuses maneuver.

5.2.3 Lasonen-Aarnio on Excuses and Dispositions

Lasonen-Aarnio's project in the paper in question (Lasonen-Aarnio (forthcoming)) is to answer the "new evil demon" problem for knowledge-first epistemology. The new evil demon problem, first developed by Cohen (1984) against Reliabilism, assumes that victims of an evil demon who reason in the same way the more rational of us do (for instance, who expect the next emerald to be green after examining thousands of green emeralds, who take experience at face value when nothing indicates they shouldn't do so, etc.), are as justified in their beliefs as we are in ours.[6] The way she answers that challenge is by introducing the notion of a reasonable (or competent) belief, and then arguing that although victims of an evil demon obviously cannot comply with the knowledge norm for belief they (and their beliefs) can be reasonable in the sense defined. Moreover, although a belief can be reasonable without amounting to knowledge, there is a conceptual connection between reasonableness and knowledge, for a belief is reasonable just in case (roughly) it is the outcome of a disposition to know. Briefly put, Lasonen-Aarnio's view is that, in believing as they do, good victims of an evil demon instantiate an extrinsic disposition to know, but this disposition is masked by the evil demon and so it doesn't result in knowledge. In this section I argue against that view, but to do so I first need to explain the metaphysics of dispositions that underlies Lasonen-Aarnio's discussion.[7]

First we need to introduce a distinction between intrinsic and extrinsic dispositions, and then explain how dispositions can be *masked* and *finked*. Dispositions are conditional-looking properties. Thus, fragility is the disposition to break when subject to appropriate forces, solubility in water is the disposition to dissolve when immersed in water, etc. Going along with their close connection to some kind of counterfactuals, a distinction is usually assumed between the *stimulus*

[6] Of course, there may be victims of an evil demon who are not like that: some of them may, for instance, expect the next emerald to be blue after examining a few thousand green emeralds. We are not interested in those victims.

[7] Lasonen-Aarnio's position is thus a hybrid of the two kinds of damage-control maneuvers I distinguished in the introduction. She holds that the victims of evil demons can have reasonable, and not merely excusable, beliefs, and thus it is not a pure version of the excuses maneuver. But neither is it a pure version of explaining rationality in terms of knowledge, for Lasonen-Aarnio doesn't commit to saying that reasonable belief just is rational belief.

conditions of a disposition and its *manifestation*. For instance, the stimulus condition of fragility is being subject to forces of the appropriate kind, and its manifestation condition is breaking, whereas the stimulus condition of solubility in water is to be immersed in water, and its manifestation dissolving.[8] Fragility and solubility in water are examples of *intrinsic dispositions*: any internal duplicate of a fragile object will be fragile, and any internal duplicate of something that is soluble in water will also be soluble in water. Some philosophers have posited the existence of *extrinsic* dispositions: dispositions that need not be shared by intrinsic duplicates. For instance, consider the property of *weighing 200 pounds* (this is an example that McKitrick (2003) takes from Yablo (1999)). This is arguably a dispositional property, the property that an object has when it is disposed to depress a properly constructed scale so as to elicit a reading of 200 pounds in the gravitational field of the object. Obviously, however, it is not an intrinsic property: a molecule-by-molecule duplicate of an object that weighs 200 pounds on Earth does not weigh 200 pounds on the Moon.[9]

A disposition is *masked* when the process that leads from the stimulus to the manifestation condition is interfered with so that the disposition is not manifested. For instance, a fragile vase may be carefully wrapped in packing material so it doesn't break when subject to forces that would have broken it had it not been so wrapped. The vase remains fragile, but its disposition to break is masked by the packing material. By contrast, a disposition is *finked* when the would-be stimulus conditions themselves bring about the object's acquisition or loss of a disposition. Consider, for instance, the following example by Martin (1994). Say that an electrical wire is *live* if and only if it is disposed to conduct electricity when touched by a conductor. Suppose now that a dead wire (i.e., one which is not disposed to conduct electricity when touched by a conductor) is connected to an electro-fink, a device that makes the wire go live whenever it is touched by a conductor. The wire is dead, but it would become live were it to be touched by a conductor—i.e., it would become live were the stimulus condition of the disposition to obtain. Similarly, consider a reverse-fink, a device that kills a live wire when it is touched by a conductor. The wire is indeed live, and yet it would lose this disposition if the stimulus conditions were present. The important difference between this case and the case of the fragile vase wrapped in packing material is that the vase remains fragile, whereas the live wire goes dead, when the corresponding stimulus condition is present.

[8] It shouldn't be assumed, however, that it is always easy to identify the stimulus and manifestation conditions for each disposition. Indeed, it might well be that the ones I identified for fragility and solubility in water are not exactly right. So-called *canonical* dispositions are explicit about their stimulus and manifestation conditions. For instance, corresponding to the non-canonical property of fragility there is the canonical disposition to break when subject to appropriate forces.

[9] A terminological problem arises here, for "pounds" is ambiguous between mass and weight. Strictly speaking, having a mass of, say, 77 kg corresponds to a weight of 756 Newtons on Earth, or 125 Newtons on the Moon. In what follows I switch to this terminology.

With these preliminaries out of the way, we can state Lasonen-Aarnio's theory of reasonability. According to a rough version of her view, "a belief is reasonable just in case it exemplifies virtues that are conducive to epistemic goods like true belief, knowledge, or belief that is proportioned to the evidence." One thing is not like the others on this list, however. The norm to believe in proportion to your evidence is much friendlier to the claim that even victims of an evil demon can have justified beliefs than the truth or the knowledge norm. Given an externalist account of evidence, such as Factualism, even this norm will in some cases conflict with the intuitions elicited by the evil demon case,[10] but there will be other cases where it doesn't—for instance, a subject who knows that the first few thousand emeralds are green will satisfy this norm if he believes that the next one will be green as well, even if he is the victim of a demon who wants to deceive him just locally about the color of emeralds. Because it will make the need for a theory of reasonableness clearer, and because it is Lasonen-Aarnio's own preferred view, I will stick to the idea that a belief is reasonable just in case it exemplifies virtues that are conducive to knowledge.

Lasonen-Aarnio thinks that the virtues conducive to knowledge are a set of dispositions to believe which regularly result in the subject's knowing that which he believes. These dispositions must clearly be extrinsic. Two subjects may be molecule-by-molecule duplicates of each other and yet one may know propositions that the other one doesn't. To begin with, as Lasonen-Aarnio reminds us, many philosophers think that the capacity to form contentful mental states at all might well be extrinsic. For instance, many philosophers think that a swamp creature formed by a lightning strike who lasts all of two seconds cannot have mental states, even if during those two seconds he is an exact duplicate of Donald Davidson when he was thirty years old. But even setting those concerns aside, a duplicate of mine may fail to know that all emeralds are green simply because he lives in a world where not all emeralds are green. The truth-condition on knowledge guarantees that a disposition to know is an extrinsic disposition. This means that Lasonen-Aarnio has her work cut out for her, for the mere fact that a victim of an evil demon is an intrinsic duplicate of mine does not guarantee that he will have all of my extrinsic dispositions. Indeed, it seems obvious that my victimized counterpart must *lack* my dispositions to know, because most of his beliefs are false. This is just like saying that an intrinsic duplicate of someone who weighs 756 Newtons on Earth will lack this dispositional property if the duplicate lives on the Moon (although these subjects will arguably share the same mass, an important point to which I return later).

[10] This will happen in cases of basic knowledge. Suppose, for instance, that my knowledge that I have hands is basic, in the sense that it is not the result of an inference from some of my other beliefs. In that case, I have some evidence (namely, that I have hands) that my deceived counterpart lacks (even if my counterpart does have hands, he doesn't know it). Therefore, the beliefs that are proportioned to my evidence might be different than those that are proportioned to his evidence.

This is a challenge that, to her credit, Lasonen-Aarnio tackles head on. She claims that some dispositions, like the disposition to know (and presumably also weighing 756 Newtons), are extrinsic because they are *anchored*.[11] A disposition is *anchored* just in case it will be manifested in the presence of the stimulus conditions only when in a certain environment. Thus, weighing 756 Newtons was a dispositional property of Neil Armstrong, anchored to the Earth. He had this property even when on the Moon, for even there he had the disposition to elicit a reading of 756 Newtons on a scale *on Earth*. The dispositional property of weighing 756 Newtons, Lasonen-Aarnio would say, was however masked by the Moon's (lesser) mass. By contrast, an intrinsic duplicate of Armstrong born and raised on the Moon does not weigh 756 Newtons, for he does not have the disposition to elicit a reading of 756 Newtons *on the Moon*. Similarly, Lasonen-Aarnio wants to say, we anchor the doxastic dispositions of victims of evil demons to environment like ours. Thus, they do have and exercise knowledge-conducive dispositions, for the beliefs they form, and the way in which they form them, would amount to knowledge in an environment like ours. It is an interesting question why we anchor the dispositions of victims of evil demons to environments like ours, and Lasonen-Aarnio spends some time explaining when we are and when we are not willing to do that. Although I think that her discussion in this respect is far from conclusive, I will grant to her that we do anchor their dispositions in that way. The claim, then, is that victims of an evil demon have dispositions to know (because they have dispositions to form beliefs that would amount to knowledge in environments like ours), but the exercise of those dispositions is masked by the evil demon, just like the exercise of the disposition to elicit a reading of 756 Newtons on a scale was masked by the Moon's gravity for Armstrong.

I find several problems with Lasonen-Aarnio's claim that victims of an evil demon have reasonable beliefs because their beliefs result from masked dispositions to know. First, her introduction of the notion of anchored dispositions seems ad hoc. Recall that, as defined by Yablo, the property of weighing 756 Newtons is the dispositional property of eliciting a reading of 756 Newtons on a scale *in the gravitational field of the object*. Thus, Armstrong has this property while on Earth, and he lacks it while on the Moon. The natural thing to say, then, is that victims of an evil demon have doxastic dispositions which would result in knowledge in environments like ours, but not in environments like theirs. We can, if we want, obfuscate these distinctions, and claim that to say that Armstrong weighs 756 Newtons is to say that he has the relevant dispositional property when on Earth, and also claim that to say that victims of an evil demon have dispositions to know is to say that they have dispositions to form beliefs which would amount to

[11] As I just said, dispositions to know are obviously extrinsic because of the truth-condition on knowledge. Presumably, Lasonen-Aarnio thinks that their anchored nature is a *further* reason why they are extrinsic.

knowledge when in environments like ours. But we should be cleared-eyed about the fact that this is what we are doing.

Both in the weight and in the doxastic case, there are five properties going around, all of them arguably dispositional, three clearly intrinsic, one clearly extrinsic, and one (Lasonen-Aarnio's anchored disposition) that in some respects behaves like an extrinsic disposition and in some other respects behaves like an intrinsic disposition. Take Armstrong's weight first. Here are the five properties in question:

1w. Having a mass of 77 kg.
2w. Weighing 756 Newtons = being disposed to elicit a reading of 756 Newtons on a scale in the subject's gravitational field.
3w. Weighing 756 Newtons on Earth = being disposed to elicit a reading of 756 Newtons on a scale on Earth.
4w. Weighing 756 Newtons on the Moon = being disposed to elicit a reading of 756 Newtons on a scale on the Moon.
5w. Weighing 756 anchored Newtons = being disposed to elicit a reading of 756 Newtons on a scale in the subject's anchored environment.

Property 1w is intrinsic (although arguably still dispositional). Property 2w is extrinsic. Two perfect intrinsic duplicates might differ regarding property 2w because they are on different gravitational fields. Properties 3w and 4w are intrinsic. Two perfect internal duplicates will agree on their distribution of properties 3w and 4w. Property 5w might seem just as extrinsic as property 2w. After all, two internal duplicates might differ with respect to their anchored environments, and thus they might also differ with respect to property 5w. So, in that respect, property 5w is extrinsic. But, as Lasonen-Aarnio puts it, it is "more intrinsic" than property 2w. Property 2w comes and goes with a change of environment. Armstrong has it while on Earth, he loses it while on the Moon. Not so with property 5w. Assuming that Armstrong is anchored to Earth, he retains property 5w while on the Moon.

Consider now the five corresponding doxastic properties (and assume that, in every case, the subjects have observed a sufficiently high number of green emeralds):

1d. Believing that all emeralds are green.
2d. Having knowledgeable dispositions = being disposed to know that all emeralds are green in the subject's environment.
3d. Having dispositions that are knowledgeable on Earth = being disposed to know that all emeralds are green on Earth.
4d. Having dispositions that are knowledgeable on Evil Earth = being disposed to know that all emeralds are green in the evil demon's environment.
5d. Having dispositions that are anchored-knowledgeable = being disposed to know that all emeralds are green in the subject's anchored environment.

The second problem with Lasonen-Aarnio's introduction of anchored dispositions is that it seems clearly false to say that they are masked when exercised in environments different from their anchoring environments. Take weight first. Disposition 1w is obviously not masked while Armstrong is on the Moon and, moreover, Armstrong has that disposition. Nor are dispositions 2w through 4w masked. Armstrong simply lacks disposition 2w while on the Moon. He also lacks property 4w, wherever he is. But he has disposition 3w on the Moon. True, this disposition cannot be manifested, but this is not because it is masked. For a disposition to be masked, remember, is for its manifestation conditions to fail to be satisfied when its stimulus conditions are present. This is not what happens with property 3w. It's not that the stimulus conditions are satisfied but the appropriate scale does not elicit a reading of 756 Newtons. Rather, what happens is that the stimulus conditions cannot obtain on the Moon, for the stimulus conditions include being on the Earth. What about property 5w? Well, given that Armstrong is anchored to the Earth, that property is masked if and only if property 3w is masked. Given that property 3w is not masked, neither is property 5w.

Analogous claims go with respect to properties 1d through 5d. Property 1d is obviously not masked for victims of an evil demon. Indeed, good victims have this property. Nor are properties 2d through 4d masked. Victims of an evil demon simply lack property 2d while being victimized. They also lack property 4d—and, indeed, arguably property 4d cannot be instantiated (the evil demon's environment is simply too hostile to make knowledge possible). They do have property 3d. True, this disposition cannot be manifested while on Evil Earth. But this is not because the disposition is masked. It's not that the stimulus conditions for property 3d are satisfied but the resulting beliefs simply do not amount to knowledge. Rather, what happens is that the stimulus conditions include being in environments like ours. What about property 5d? Well, given that (following Lasonen-Aarnio) we are assuming that the victims are anchored to Earth, that property is masked if and only if property 3d is masked. Given that property 3d is not masked, neither is property 5d.

It is at the very least not clear, then, that Lasonen-Aarnio's anchored dispositions are masked. But regardless of whether she is right that they are masked or I am right that they are not even triggered, they are clearly not manifested. Obviously, no belief can be the manifestation of a disposition that is not manifested. Lasonen-Aarnio thinks that by appealing to anchored dispositions she has successfully defended the claim that victims of an evil demon can have and exercise dispositions to know. But notice that for a belief to amount to knowledge, it must be the *result* of a correct exercise of a knowledgeable disposition. If a doxastic disposition is not manifested, then of course it cannot be responsible for any belief's amounting to knowledge. Similarly, for a belief to be reasonable, it must be the result of the exercise of a disposition to know. But, again, a disposition that is not manifested (whether because it is masked or because it is not triggered) does not result in any belief. Simply *having* the disposition *while* one forms a belief does not mean that the disposition produced the belief. Armstrong has the disposition to elicit a reading of 756 Newtons on a scale on Earth, and while he has that disposition he elicits a reading of 125 Newtons on a scale on the Moon. This elicitation is in no way a manifestation of, or the exercise of, or an exemplification of, his disposition to elicit a reading of 756 Newtons on a scale on Earth.[12] Similarly, a victim may well have the disposition to know that all emeralds are green when in a normal environment, and, while having this disposition, he can believe that all emeralds are green. This belief is in no way a manifestation of his disposition to know that all emeralds are green in an environment like ours.

5.3 Justification as Would-Be Knowledge

Lasonen-Aarnio has therefore not successfully argued that the beliefs of victims of an evil demon are reasonable, in her sense of "reasonable." But there is an approximately true claim in the vicinity of Lasonen-Aarnio's false claim. The false claim is that the beliefs of victims are the results of dispositions to know in environments like ours. This is false because no belief is the result of that disposition on Evil Earth. The approximately true claim in the vicinity is that the beliefs of victims are the results of dispositions that, were they to be exercised in our environment, would result in knowledge.[13] Or, to put it in terms more friendly to Lasonen-Aarnio, they are produced by dispositions that, were they to be exercised in the subject's anchoring environment, would result in knowledge. What if we take "reasonable" to mean this instead?

This proposal is in line with a couple of views which fall under the second form of damage control mentioned in the introduction to this chapter. This second form of damage control, remember, is more concessive than the excuses maneuver as implemented by either Williamson or Lasonen-Aarnio, for it grants that false beliefs can be justified or rational. However, it still counts as being in the knowledge-first camp because it claims that rational belief can be defined in terms of knowledge. I turn now to an examination of such views, and later I connect them with the proposed revision to Lasonen-Aarnio's view.

The idea of defining justification in terms of knowledge was first proposed, as far as I can tell, by Sutton (2005). There, Sutton says:

> We only understand what it is to be justified in the appropriate sense because we understand what it is to know, and we extend the notion of justification to

[12] All of that is compatible, of course, with both dispositions having a common explanation, in terms of Armstrong's mass. More on this in section 5.3.

[13] I say "approximate truth" to inoculate the claim against the conditional fallacy.

> non-knowers only because they are would-be knowers. We grasp the circumstances—ordinary rather than extraordinary—in which the justified would know. Justification in the relevant sense is perhaps a disjunctive concept—it is knowledge or would-be knowledge. (Sutton (2005), p. 10)

Bird (2007) takes Sutton's suggestion and codifies it as follows:

JuJu: If in world w_1 S has mental states M and then forms a judgment, that judgment is justified if and only if there is some world w_2 where, with the same mental states M, S forms a corresponding judgment and that judgment yields knowledge.

Bird means JuJu to be understood diachronically. A synchronic version of JuJu would amount to saying that S's belief is justified just in case someone with the same mental states as S could know. But Bird wants to follow Williamson in counting knowledge as a mental state, and so a synchronic version of JuJu would just collapse justification with knowledge—something that both Sutton and Williamson are friendly to (see Sutton (2005) and Williamson (forthcoming)), but that Bird wants to avoid. I don't think it's obvious that thinking of JuJu as diachronic will help Bird with this issue, for there may well be cases where the justified belief is formed contemporaneously with the relevant mental state M, but I won't fuss about that issue here.

Ichikawa (2014) proposes a similar account of justification:

JPK: S's belief is justified iff S has a possible counterpart, alike to S in all relevant intrinsic respects, whose corresponding belief is knowledge.

Ichikawa's definition sidesteps the issues having to do with counting knowledge as a mental state, for it compares S's belief with a counterpart which is an intrinsic, not a mental, counterpart. Even Williamsonians who claim that knowledge is a mental state grant that it is not an intrinsic state—as I already mentioned, the truth-condition knowledge alone suffices to make it an extrinsic state.

All of these proposals (including the modified Lasonen-Aarnio view) must face the following issue: it may well be that we can be justified in believing unknowable propositions. I offer three examples. Take, first, the kind of proposition at issue in the so-called "knowability paradox" due to Church.[14] For concreteness's sake, let us suppose that I will never taste Marmite again in my life, but that I do not know it. That conjunctive proposition cannot be known by me. For suppose that I do

[14] The "paradox" first appeared in Fitch (1963), where Fitch attributed it to an anonymous reviewer. Joe Salerno discovered that the anonymous reviewer was Alonzo Church—see Salerno (2009). I use scare quotes because I agree with the prevailing view that the result is not at all paradoxical.

know it. Then I know both that I will never taste Marmite again in my life and that I do not know that I will never taste Marmite again in my life. Given the factivity of knowledge, however, I do not know that I will never taste Marmite again in my life. Contradiction.[15] But although I cannot know those kinds of propositions, cannot I be justified in believing them? Suppose that I have two friends in the Marmite industry. One of them tells me that I will never taste Marmite again, the other that I don't know that what the first one told me is true. Can't I trust them both? *One* reason why my second friend tells me what he does may be that they disagree on the first-order issue of whether I will ever taste Marmite again. But that need not be the only reason, and I need have no opinion whatsoever about what their reasons are for telling me what they do—I just straightforwardly trust them.

Another example along similar lines is afforded by lottery propositions. A standard take on them is that whereas I cannot know that my ticket will lose based solely on the fact that it is a ticket on a fair lottery, I can be justified in believing it. But this possibility is closed off by definitions of justification as would-be knowledge.

Unknowable propositions and lottery cases are tricky in many ways, so they do not make for the best counterexamples. But I take the following third example to be pretty damaging for theories of justification as would-be knowledge. Suppose that I am convinced by skeptical arguments and think that knowledge is impossible. I may, of course, be wrong—but can't I be justified in believing that knowledge is impossible? It seems undeniable that I can. And yet, it is impossible to know that knowledge is impossible. Therefore, theories of justification as would-be knowledge must deny that it is possible to be justified in believing that knowledge is impossible.

There are powerful reasons to think, then, that theories of justification as would-be knowledge are materially inadequate. But in the remainder of this section I want to set that problem aside, to concentrate on other issues that affect them. Let us suppose, then, that some such theory is materially adequate. What then? Why would it be a vindication of knowledge-first epistemology? In supposing for the sake of argument that some account of justification as would-be knowledge is materially adequate we are only assuming that some biconditional linking justification to possible knowledge is true. This by itself, of course, does not amount to a vindication of knowledge-first epistemology. It may well be that justification is would-be knowledge (in the sense that some biconditional of the sort just mentioned is true) and yet that justification can be understood independently of knowledge. Indeed, some such biconditional can be true while at the same time knowledge must be explained in terms of justification—as anti-

[15] Factivity is actually not required to generate the result. The result applies to any operator O that distributes over conjunction and that validates the inference from $O\neg O(p)$ to $\neg O(p)$.

knowedge-first a claim as they come. What knowledge-firsters must have in mind when offering their biconditionals, then, is not only that they are true, but also that there is an explanatory relation underlying their truth, an explanatory relation that goes from the knowledge side of the conditional to the justification side. How plausible is this further assumption? Not very plausible, I think. One way to see the implausibility of the idea that justification can be explained in terms of knowledge is to go back to the modification to Lasonen-Aarnio's theory I suggested at the beginning of this section.

That modification, recall, left us with the following claim: the beliefs of victims of an evil demon are reasonable because they are formed by dispositions that would lead to knowledge if activated in our environment. But notice that the dispositions in question—the ones that lead to false beliefs in the evil demon scenario, but would lead to knowledge in environments like ours—are wholly intrinsic. Indeed, to put it in terms that are not friendly to knowledge-first epistemology, they are simply dispositions to form rational beliefs.

What Lasonen-Aarnio has in mind when she says that anchored dispositions to know are extrinsic is that two intrinsic duplicates need not share them, because they may be anchored to environments that require different dispositions to produce the same result. Thus, she gives the example of

> a world in which light reflected from objects is distorted in a systematic manner. Having evolved in the light-distorting environment, the alien cognizers inhabiting it have a visual apparatus that takes the distortions into account. As a result, their visual apparatus is disposed to represent the world fairly accurately. Were they to land on Earth, they would not have accurate perceptual experiences and would, let us assume, form false beliefs about their surroundings. But this seems irrelevant for evaluating whether they are disposed to form accurate beliefs based on their visual experiences—just as the fact that there are pockets of the Universe where we would be prone to form highly inaccurate beliefs seems irrelevant when evaluating our dispositions to acquire true beliefs and knowledge.

This passage echoes a case I had considered in Comesaña (2002), a case of

> subjects who inhabit a world where light travels in funny ways, so that, e.g., when they "see" something in front of them, it is really to one side, but are so constituted that when they "see" something in front of them they form the belief that it is to one side. (Comesaña (2002), p. 262)

There seems to be a difference between the way I was conceiving of the scenario and the way Lasonen-Aarnio does. I was conceiving of the scenario as follows: whenever there is something slightly to the right of these subjects, their experience represents it as being right in front of them, but they believe that it is slightly to their right. The quote from Lasonen-Aarnio suggests a different interpretation: whenever there is something slightly to the right of these subjects, their experience correctly represents it as being slightly to their right, and they form beliefs accordingly. In my version, the subjects correct for the distorting light at the doxastic level, whereas in Lasonen-Aarnio's version the correction comes at the experiential level already. I don't think that on either case of construing the case are the relevant dispositions extrinsic.

Take first my way of construing the case. Things are different depending on how the details of the case are filled in, but either way the subjects do not have a disposition to know anchored to their environment. On one way of filling in the case, the subjects can discover, through the normal course of scientific research, that light behaves in this strange way in their environment. Thus, these subjects start out their doxastic life just like ourselves, believing that objects that seem to be in front of them really are in front of them. Later, they learn that light behaves in funny ways, and that objects are really slightly to one side when they seem to be right in front. Despite appearances, these subjects simply have the same dispositions we have. We too would believe that objects that seem to be right in front of us are actually to one side, were we to learn that light behaves in that way. And they too would have continued to believe that objects that seem to be right in front of them are right in front of them, were they not to have learned that light behaves that way. This is simply a case where these subjects have different evidence than we have, not a case where they respond differently to the evidence they share with us. So that way of filling in the case simply does not make the case that Lasonen-Aarnio thinks it does.

On the alternative, more radical way of filling in the case, there is simply no way to find out that light behaves in that strange way. This is, in effect, an evil demon kind of scenario, where the deception is global and systematic. And here is the crucial claim: subjects who, in this situation, form true beliefs, far from displaying virtues that are conducive to knowledge in their environment simply have irrational beliefs in conspiracy theories. They believe that they are radically and systematically misled about the location of objects in their environment. The fact that they are indeed radically and systematically misled does nothing to remove their irrationality. Just because they are after you doesn't mean you aren't paranoid. Victims of an evil demon cannot know or rationally believe that they are victims of an evil demon. Regardless of how well adapted to their environment these alien cognizers are, they do not have knowledge.[16]

What this shows is that some worlds are simply too hostile to allow for knowledge. So, what does it take for a world *not* to be so hostile? A somewhat parochial answer is that it must be sufficiently like our world, where believing in

[16] The phrase "alien cognizers," and the idea behind it, is from Bergmann (2006).

the way we do suffices for knowledge in many cases. But this *is* parochial. Our knowledge is due to the way we form and maintain beliefs as much as it is due to the world's being the way it is. Indeed, we should not forget that, in addition to our having a lot of knowledge, we are also vastly ignorant—although it is in the nature of the case that we cannot identify that ignorance. Some of that ignorance can be identified in retrospect. But some may be irremediable, as Kant and some contemporary mysterians would have it. Indeed, some of our ignorance may be of a particularly insidious kind, the truth of the matter getting farther and farther away the more we know. If we were, by blind luck, to have dispositions to have true beliefs about such matters, we would still not know them. So: while it is true enough that for a world to allow for knowledge it must be sufficiently like ours, the order of explanation is reversed from the one required by knowledge-firsters. Our world allows for knowledge because believing rationally leads to believing truly, not because we have whatever dispositions we need to have in order to believe truly in our world. This means that dispositions to know are invariant with respect to environments.[17] If an environment does not allow for knowledge using the dispositions that give rise to knowledge around here, then it does not allow for knowledge, period.

Now take Lasonen-Aarnio's construction of the case. On that interpretation, the subjects' experiences are well matched to their environment, and they accurately represent the positions of objects in their environment. This is just like the first way of filling in my interpretation of the case in the important respects. They believe, just like we do, that objects are wherever our experience represents them as being. True, were we to be in their environment, we would have misleading evidence, and were they to be in our environment, they would have misleading evidence. But what does that show? I said before, in examining my interpretation of the case, that some worlds are too hostile to allow for knowledge, for instantiating the intrinsic dispositions which are necessary for knowledge would not lead to true beliefs in those worlds. Lasonen-Aarnio's interpretation of the case highlights the fact that hostility can be relative. We do not (and cannot, given our sensory apparatus) instantiate knowledgeable dispositions in the alien's environment, and the aliens do not (and cannot, given their sensory apparatus) instantiate knowledgeable dispositions in our environment. Still, in order to know (in either environment) the right intrinsic disposition needs to be instantiated. In the case in question, this is the disposition (roughly) to take experience at face value.

What does this mean for Lasonen-Aarnio's project of explaining the intuitions behind the new evil demon objection in a knowledge-first-friendly way, and for accounts of justification as would-be knowledge? It means, I think, that these projects fail. If, as I've argued, knowledge requires a fixed set of dispositions, then

[17] Or, at least, invariant with respect to the environment outside the subject. One can hold that rationality is subject-relative in that, e.g., it depends on the subject's cognitive capacities.

this set comes first, in the sense that we can theorize about it independently of knowledge. We can then agree nominally with Lasonen-Aarnio, at least when her thesis is restated the way I suggested: reasonable beliefs are those that are formed by dispositions that would result in knowledge in the anchoring environment of victims of an evil demon (assuming that their anchoring environment is our environment). We can also agree with Bird or Ichikawa that some biconditional linking justification and knowledge is true (at least if we assume that the counterexamples offered earlier don't work). But this is because an environment is friendly to knowledge only if believing in accordance with that fixed set of dispositions results in beliefs that are true in that environment. This is simply a roundabout and not particularly illuminating way of specifying the fixed set of dispositions that are necessary for knowledge. There is another, more traditional way of identifying that set: it is the set of dispositions that produces justified beliefs.

My example of alien cognizers concerns what is arguably a case of basic belief-formation: the beliefs that an object is right in front of me, or that it is to my left, are supposed to be formed as immediate reactions to experience, and not on the basis of any other beliefs. But analogous claims apply to cases of inferential belief. In normal environments like ours, we can come to know that the next observed emerald will be green on the basis of observing a sufficient number of green emeralds under suitable conditions. The possibility of acquiring inferential beliefs like this one is accompanied by two features of the ur-prior: the fact that the conditional probability of emerald $i + 1$ is green given that emeralds 1 through i are green is (a) higher than its unconditional probability; and (b) sufficiently high in absolute terms. Feature (a) corresponds to the fact that the belief that emerald $i + 1$ is green is based on the previous beliefs that emeralds 1 through i are green, and feature (b) corresponds to the fact that my degree of confidence in the proposition is high enough to give rise to full belief. I emphasize that I am neutral regarding the relationship between credences and beliefs. I do not, in particular, either endorse or reject a reduction of belief to credence or vice versa. Still, I take it to be a very plausible constraint on any view of the relationship between credence and belief (whether reductionist or not) that low credence cannot be accompanied by belief, at least not in these paradigmatic cases of inferential belief-formation.

Now suppose that a subject inhabits a world where there are lots of emeralds. The vast majority of those emeralds are green, but a tiny minority are blue. The subject in question adheres to a perverse ur-prior, according to which the credence that emerald i is green gets lower and lower when conditionalizing on each belief of the form *emerald n is green*, for n between 1 and $i - 1$. Suppose that, as luck would have it, emerald i is one of the blue ones. The subject, then, has ever higher credence in the truth regarding the color of emerald i the more evidence he acquires. Indeed, we can assume that by the time he gets to observe emerald $i - 1$ he all out believes that emerald i is not green. As before, however, this belief counts not as knowledge but as lucky superstition. It's not that we can infer that the next emerald will be green based on a sufficient number of observed green emeralds because that is the kind of disposition that is conducive to knowledge around here; rather, we can know that the next emerald will be green because our environment cooperates and makes it the case that true belief often enough accompanies rational belief. Rational inferences conform to the uniquely rational ur-prior, and knowledge is possible only in environments where the world cooperates and conforms to the ur-prior as well (in the sense of making true propositions which receive sufficiently high credence given the ur-prior and the requirement of ur-prior Conditionalization).[18]

We can try to explain this primacy of justified belief over knowledge by appealing to our judgments about certain counterfactuals. Thus, we tend to assent to counterfactuals such as the following: even if the world had been different, the same beliefs would have been justified given the same evidence; in particular, even if we had been victims of an evil demon, the same beliefs that are justified for us now would be justified for us then. Excuses moves, such as Williamson's and Lasonen-Aarnio's, seek to undermine the probative value of such counterfactuals by claiming that such judgments track, not the notion of rationality, but some alternative notion—excusability in Williamson's case, reasonability in Lasonen-Aarnio's. Accounts of justification as would-be knowledge are more concessive, in that they grant that those judgments do track genuine rationality, but they claim that the notion of rationality in question can be explained in terms of knowledge. Thus, those accounts too seek to undermine the value of our judgments about those counterfactuals as arguments against knowledge-first epistemology.

Indeed, my own view in Comesaña (2002), cited above, could be adapted for this job. In that paper I defended Reliabilism from Cohen's new evil demon objection. My view there was that we should think of justification as requiring reliability in the actual world. Given a two-dimensional semantics for "actual," this means that ascriptions of justification express two propositions: the proposition that the belief in question was produced by processes that are reliable in whichever world the belief inhabits, and the proposition that the belief in question was produced by processes that are reliable in the speaker's world.[19] Thus, our counterfactual judgments do track Cohen's intuitions about evil demon cases, but they do not undermine Reliabilism. This view can be adapted to the knowledge-first program. Simply claim that beliefs are justified just in case they are produced in a way that would result in actual knowledge.

[18] Notice that, interestingly, there is no room in this case for an interpretation like Lasonen-Aarnio's interpretation of the alien cognizer. In the case of inferential knowledge, there is no third factor relevant to knowledge, aside from the intrinsic disposition and the truth of the belief in question (I'm setting aside Gettier conditions, which arise for the basic case as well).

[19] But cf. Ball and Bloome-Tillmann (2013). For the record, my view at the time was the one they mention on p. 1327.

Given these views, we cannot explain the primacy of justification over knowledge simply by appealing to our judgments about those counterfactuals, for views according to which there is no such primacy can predict those judgments. Maybe fancier counterfactuals could do the job. But we don't need them. The view is not just *that* those counterfactuals are true, but includes a thesis about *why* they are true. That thesis is that whether a belief is rational is an intrinsic property of the processes by which the belief was acquired and sustained. This allows for a *measure* of extrinsicness, insofar as it opens the door for views that do not conform to the strictures of time-slice epistemology. According to time-slice epistemology, justification supervenes on instantaneous intrinsic states. Thus, according to time-slice epistemology, if subject S at time t_S is intrinsically identical to subject S' at time $t_{S'}$, then S and S' are exactly alike as far as justified beliefs go (i.e., for any proposition p and doxastic attitude D, S is justified in adopting D towards p if and only if S' is justified in adopting D towards p). I am sympathetic to time-slice epistemology, but I will not defend it here. As I said, my formulation of the thesis about what it is, exactly, that underlies our judgments about the evil demon counterfactuals is compatible with the denial of that view. It may be that justification does not supervene on instantaneous intrinsic states, but the claim is that it supervenes at least on the state and history of the belief in question. In particular, no mention of knowledge (or reliability) is needed to explain why the counterfactuals in question are true.

I am not denying that the excuses maneuver, or accounts of justification as would-be knowledge, can give knowledge-firsters what *they* want. But there is a cost to all of these defenses of knowledge-first. The cost is the claim that, in an important sense, excusability or justification are contingent. I say "in an important sense" because the views in question can mimic necessity. My own view, for instance, had it that one of the propositions expressed by attributions of justification was necessary. But the necessity in question is not deep in a sense that is easy to grasp but hard to express precisely. One way to bring out the relatively shallow nature of the necessity in question is to appeal, once again, to environments that are unfriendly to the development of knowledge as we have it here. According to all of these theories, subjects with sufficiently perverse belief-forming dispositions count as having rational beliefs (in the view I proposed, there is one sense of reliability in which the beliefs of these subjects count as reliable). But they *are* perverse, not rational.

Thus, even if knowledge-firsters can appease their own consciousness with defensive maneuvers or error theories, there is a limit to what they can achieve. In particular, they will not be able to capture the deep necessity of norms of rationality in a way that allows them to explain why rational belief is still possible in environments where knowledge is unattainable.

Another analogy with weight may help here as well. There is an intrinsic property of Armstrong, namely his mass, which explains why he weighs what he does in different environments. It is because he had a mass of 77 kg that he weighed what he did on Earth and the Moon. It is because he had that mass that, had he been to Mars, he would have had some different weight. With sufficient technical sophistication we could, perhaps, develop weight-first theories which mimic many of the counterfactuals entailed by this primacy of mass over weight. But weight-first theories can at best give us pale and superficial imitations of what is in this case the obviously true mass-first theory. This is how knowledge-first theories look to us advocates of rationality-first.

5.4 Conclusion: Rationality-First

If my arguments of the previous chapter and this one are on the right track, then we should abandon both knowledge-based decision theory as well as knowledge-first epistemology. What should be put in their place? Rationality-based decision theory and rationality-first epistemology. The arguments of chapter 4 against allowing inferential knowledge to play the role of evidence apply equally well against allowing inferential rational belief to play that role. Thus, rationality-based decision theory is the combination of the following claims: rational actions are those that maximize expected utility relative to rational beliefs and credences; rational credences are those that are the result of conditionalizing the unique ur-prior on whatever evidence the subject has at the relevant time; the evidence a subject has at a time is constituted by the beliefs that the subject is basically (non-inferentially) rational in believing at that time. In line with the arguments of this and the previous chapter, the view has it that it is possible to be basically justified in believing false propositions—for instance, Tomás is justified in believing that Lucas is offering candy in Bad Lucas. In chapter 6 I turn to the development of a theory of basic justification which has precisely this consequence.

6
Experientialism

6.1 Introduction

In the previous two chapters I argued for the conclusion that we can rationally believe false propositions. The main assumptions of that argument are that we can rationally act on the basis of false propositions and that rational action requires rational belief. The picture I ended up with can be put in different frameworks. In terms of the objective Bayesian framework, there is a unique evidential probability function Cr_u in whose terms we can define a conditional evidential probability function $Cr_u(-|-)$. I take the rational credences for a subject with evidence E to be those determined by $Cr_u(-|E)$. This does not amount to the Bayesian requirement of Conditionalization, because it allows that rational subjects may lose as well as gain evidence. I argued that the evidence a subject has at a time is given not by whatever the subject believes at that time (pace Parfit) nor by what the subject knows at that time (pace Williamson), but rather by what the subject is basically rational in believing at that time.[1] The same picture can be put in terms that are more familiar in mainstream, coarse-grained epistemology. I assume that there is a unique set of epistemic principles, which dictates which attitude (belief, disbelief, suspension of judgment) is rational for a subject to assign to any proposition given a body of evidence. These epistemic principles replace the evidential probability function, and they determine not a continuous range of doxastic attitudes but only the traditional three. Other than that, however, the pictures are interchangeable. Both the evidential probability function as well as the set of epistemic principles can be seen as structuring an objective relation of epistemic support which holds between propositions. A subject will be *ex ante* rational in assigning the corresponding doxastic attitude just in case he *possesses* the reasons (or evidence) in question, and his attitude will be *ex post* rational just in case it bears the right relation to that possessed evidence. As before, then, we have (at least) three possible views regarding what evidence a subject has at a

[1] Recall that, for Parfit himself, there is a difference here between the rationality of beliefs and the rationality of actions, in that beliefs based on other beliefs do require those prior beliefs to be rational, whereas for an action to be rational it need not be based on a rational belief. It is not clear, therefore, that Parfit himself can have a coherent view of evidence. When talking about inferential beliefs, evidence needs to be given by rational beliefs, whereas when talking about action evidence can be given by irrational beliefs. In what follows I ignore this idiosyncrasy and continue talking of the "Parfitian" view according to which neither rational beliefs nor rational actions require rational belief.

Being Rational and Being Right. Juan Comesaña, Oxford University Press (2020). © Juan Comesaña.
DOI: 10.1093/oso/9780198847717.001.0001

time: the Parfitian view, according to which the subject's evidence is constituted by those propositions he believes at the time; the Williamsonian view, according to which the subject's evidence is constituted by those propositions he knows at the time; and my view, according to which the subject's evidence is constituted by those propositions he is *ex ante* basically justified in believing at the time. In this chapter I set aside the Parfitian view, and consider the Williamsonian view (aka Factualism) and my own preferred Experientialism, which respects the arguments from the previous chapters in allowing for false evidence.

Another important feature of my view is the fact that it requires *basic* rational belief as a condition on evidence-possession. The basicality constraint is motivated by the chain-reaction argument rehearsed in chapter 4: if all rational belief is allowed as evidence, including inferentially rational beliefs, then this leads to requiring evidential probability 1 for propositions that should not receive it, and this has further incorrect downstream effects on the rationality of actions. I argued that Williamson's own Factualism is subject to the same problem, so this restriction is not a feature of my theory alone. The ur-credence function, then, determines which doxastic attitudes are *inferentially* rational for a subject to adopt, whereas the theory of evidence will determine which propositions are the input to the inferential machine. This chapter is concerned with the theory of evidence.

Recall that I am assuming a conception of evidence as the set of propositions that determines which states should be considered when deciding what to do. As with the previous chapters, my arguments in this chapter will appeal to considerations regarding practical rationality to draw epistemological conclusions. My contention will be that other views fail precisely because they determine the wrong states to be considered in decision-making.

6.2 Three (or Four) Views

What is the role of experience in determining the rationality of our beliefs? One aspect of this question on which I will not spend much time is the issue of exactly which beliefs are the ones whose rationality is determined by our experience. I will not spend much time on that issue not because it is not intrinsically interesting, but rather because it is not the focus of disagreement between my view and its main competitors. I will in what follows simply assume that there are some propositions that can be the contents of experiences, and that the rationality of believing *those* propositions is somehow determined by an experience with that same content. I will, then, assume that experiences have representational content, but I shall remain neutral about phenomenal content and its relationship to representational content. The question is, how does an experience with the content that p determine the rationality of a belief that p? From time to time, I will replace the generic *p* with specific propositions, but no deep theoretical commitment should be read in those examples.

In this section I merely present three (or four) views about that question. In the sections that follow I argue for one of the views and against the others.

The first view is Factualism. According to Factualism, an experience that *p* makes a belief that *p* rational for a subject if and only if that experience provides the subject with knowledge that *p*. Thus, according to Factualism only some experiences with the content that *p* rationalize a belief that *p*. In particular, an experience with the content that *p* does not rationalize a belief that *p* when it is false—for, when *p* is false, an experience with the content that *p* cannot provide knowledge that *p*. Thus, for the Factualist, we can divide the experiences with the content that *p* into the good ones and the bad ones, according to whether they provide the subject whose experience it is with knowledge that *p* or not. All experiences that *p* had when *p* is false will thus be bad experiences in this sense (although they may be perfectly good experiences in other senses of the term).

The second view is more traditional. Williamson calls it "the Phenomenal Conception of Evidence," and I have called it, following Dancy (2000), "Psychologism."[2] According to Psychologism, an experience that *p* is itself evidence that *p* is true, and thus always provides some reason to believe that *p*. If that reason is not defeated (the defeat may happen if the subject has stronger reason to not believe that *p*, or if he has some "undercutting" defeater for *p*[3]), then the experience rationalizes a belief that *p*.

The metaphysics of Psychologism is not clear in the literature. There are several options about what basic perceptual reasons are, according to Psychologism. Suppose that I have an experience *e*. What evidence do I thereby have? Here are three salient options: *e* itself; the fact that I have *e*; my having *e*. One is an object (or, at any rate, a state), another a fact, and the third an event.

There are a number of commonalities and differences between Factualism and Psychologism. Factualism and Psychologism agree that, when experience provides us with reasons, those reasons are true. For the Factualist, the reason provided by an experience is the content of the experience, and it is provided only when that content is known (and, therefore, true). For Psychologism, the reason is the experience itself, and it is therefore always so provided when the experience exists. Thus, proponents of Psychologism and Factualists disagree about the metaphysics of basic empirical reasons—Factualists think they are known propositions, whereas proponents of Psychologism think that they are experiences—but they

[2] It is not easy to classify authors as clearly defending this position. Perhaps Conee and Feldman (1985) and Pollock (1986) come closest. Sometimes Pryor (2000) is also associated with the view, but it seems to me that Pryor's views are neutral between Psychologism and my own Experientialism, to be introduced momentarily.

[3] For this distinction between rebutting and undercutting defeaters, see Pollock (1986). I discuss defeaters later in this chapter.

agree in rejecting the idea that experience can provide us with false reasons. Another important difference between Psychologism and Factualism is that although both positions treat experience as providing non-inferential justification for beliefs, Psychologism treats experience as providing *evidential* justification. The experience itself is evidence for a belief in its content, according to Psychologism, whereas according to the Factualist it is the content itself that is the evidence. To put it in rough and ready terms, for the Factualist but not for the proponent of Psychologism it is possible for the proposition that is the content of the experience to be the first item of evidence that the subject ever receives.

Another crucial difference between Psychologism and Factualism is in their treatment of cases where a subject has an experience that *p* while the proposition that *p* is false. In that case, according to the Factualist no reason is provided for the subject to believe that *p*, so that even if he has no defeaters for the belief that *p* he will not be rational in believing that *p*.[4] According to the proponent of Psychologism, on the other hand, the experience itself is still a reason for the subject to believe that *p* even when its content is false, and so if the reason thus provided is not defeated then the subject is rational in believing that *p* even when *p* is false. Psychologism is thus friendly to the view defended in the previous two chapters about the rationality of false beliefs, whereas Factualism is incompatible with it.

The view that I favor, which I call "Experientialism," agrees with Factualism in claiming that experience provides us with reasons but is not identical to the reasons provided, but agrees with Psychologism in claiming that even bad experiences can rationalize belief.[5] One way of thinking of Experientialism is as a generalization of Factualism. Factualism holds that experiences provide their content as a reason for belief only when belief in the content counts as knowledge; Experientialism holds that experiences provide their content as a reason for belief when belief in the content is justified. Thus, Experientialism holds not only that experience can rationalize false beliefs (in this, Experientialism sides with Psychologism and against Factualism), but also that experience can provide us with *false evidence*. For Experientialism as well as for Factualism, the proposition which is the content of the experience can be the first item of evidence ever received by a subject, but for Experientialism, as opposed to Factualism, that item can be false. Thus, Experientialism agrees with Factualism about the metaphysics of reasons, but disagrees with it about the epistemology.

According to Experientialism, then, an experience provides its content as a reason when the subject is justified in believing its content. The justification in

[4] Although it could of course happen that the experience doesn't provide the subject with knowledge *because* it is defeated.

[5] Goldman (2009) proposed something like Experientialism as superior to E = K, but not on the basis of the decision-theoretic arguments to be presented here. Schroeder (2008) advocates something like Experientialism. See also Dancy (2000) and, for Matthew McGrath's and my own development of the view, Comesaña and McGrath (2014) and Comesaña and McGrath (2016).

question must be *ultima facie*. If an experience of mine provides me with *prima facie* justification for believing its content but this justification is defeated by something else I am justified in believing, then I do not have the content of the experience as evidence. This raises the obvious question: when are we *ultima facie* justified in believing the contents of our experience? Notice that an analogous question arises for Factualism. According to Factualism, an experience provides its content as a reason when the subject knows its content, which raises the obvious question: when do we know the contents of our experience? Factualists who adhere to knowledge-first epistemology may claim that this question does not merit an answer, for we cannot explain what knowledge amounts to in more basic terms. Experientialists can take this route also: they can adhere to a rationality-first program, according to which we can perfectly well tell when we are justified in believing the contents of our experience even if we cannot give a non-circular definition of empirical justification. But Experientialists can also say something more, borrowing a page from Psychologism. They can say that an experience always provides *prima facie* justification for belief in its content, and that this justification becomes *ultima facie* justification when it is not defeated. Later in this chapter I criticize Pollock's account of defeaters, in particular his view on undercutting defeaters. But this criticism does not amount to a wholesale rejection of the very idea of defeaters. Thus, it is still open to Experientialists to claim that experience provides its content as evidence when the justification it provides is not defeated. As we shall see, there are interesting issues regarding how to deal with defeaters in this account, but they are issues that arise for anyone willing to countenance defeaters.

Notice that while it is straightforward to couch Factualism and Experientialism in the Bayesian framework of the previous chapters, couching Psychologism in that framework requires some thought. According to Factualism, an empirically basic proposition gets to be part of your evidence at a time provided that it is known—and it is the proposition thus known that it is your evidence. According to Experientialism, an empirically basic proposition gets to be part of your evidence provided that it is justified by an experience you are having. So, when you have an experience with the content that p and you thereby know that p, both Factualism and Experientialism agree that it is part of your evidence that p, and so you should update your credences by conditionalizing Cr_u on p (and whatever else is part of your basic evidence at the time). But what would the proponent of Psychologism say about this? This is one place where the unclarity about the basic ontology of evidence according to Psychologism comes into play. One option is for the proponent of Psychologism to say that you should update on the proposition that you are undergoing an experience with the content that p. This option sticks most closely to the letter of Psychologism, according to which the experience itself *is* the evidence. This option has two important consequences, however. First, it leaves Psychologism open to the problem of easy knowledge. As I will argue in chapter 8, however, this is a problem that *every* theory needs to face, so this consequence is not as problematic as many have thought it to be. Second, it is natural to hold that when *E* is part of your evidence it is rational for you to believe that *E*. Indeed, part of what it means to have conditionalized on having had a certain experience is that you should have credence 1 in having had that experience. Perhaps the proponent of Psychologism will be happy with this consequence. But perhaps they will want to grant that when you have an experience with the content that *p* you need only be rational in believing that *p* itself, and not, in addition, that you had an experience with that content.

Another option for couching Psychologism in the Bayesian framework is to say that it agrees with Experientialism: when you have an experience that justifies belief in *p*, it is *p* itself that you should update on. This option doesn't have the problems of the previous one, but it does render the difference between Psychologism and Experientialism one that cannot be seen in the Bayesian framework.[6]

The views presented so far can be summarized in the following table:

	What are basic reasons in the good case about?	Are basic reasons the same in the good and the bad case?	Can basic reasons be false?
Factualism	Content	No	No
Psychologism	Experience	Yes	No
Experientialism	Content	Yes	Yes

Let me illustrate the table with the Good and Bad Lucas example. I assume that Tomás has an experience with the content that Lucas is offering a piece of candy. According to both Factualism and Experientialism, in Good Lucas Tomás's basic reason is that Lucas is offering a piece of candy. According to Psychologism, by contrast, Tomás's basic reason is that he has an experience with the content that Lucas is offering candy. Notice that, in attributing this view to Psychologism, I have forced an interpretation of it according to which basic reasons are propositions—propositions about experience—rather than experiences. Thus, in this respect Experientialism agrees with Factualism and disagrees with Psychologism. On the other hand, both Psychologism and Experientialism have it that Tomás's basic reason is the same in Good Lucas and Bad Lucas, whereas Factualism disagrees. Thus, to put it in a quick and dirty way, Experientialism sides with Factualism over Psychologism regarding the good case but sides with Psychologism

[6] For more on this issue of how to interpret views like Psychologism in a Bayesian framework, see Pryor (2013a).

over Factualism regarding the bad case. However, Experientialism is on its own when it comes to whether basic reasons themselves can be false. Both Psychologism and Factualism have it that they cannot, whereas Experientialism has it that they can.

So far, I have presented three views about basic perceptual beliefs. I will discuss those views further in the sections to follow. A fourth view is also suggested by Williamson. Considering a case of deception, he says:

> In unfavorable circumstances, one fails to gain perceptual knowledge, perhaps because things are not the way they appear to be. One does not know that things are that way, and E = K excludes the proposition that they are as evidence. Nevertheless, one still has perceptual evidence, even if the propositions it supports are false. True propositions can make a false proposition probable, as when someone is skillfully framed for a crime of which she is innocent. If perceptual evidence in the case of illusions consists of true propositions, what are they? The obvious answer is: the proposition that things appear to be that way. Of course, unless one has reason to suspect that circumstances are unfavorable, one may not consider the cautious proposition that things appear to be that way; one may consider only the unqualified proposition that they really are that way. But it does not follow that one does not know that things appear to be that way, for one knows many propositions without considering them. When one is walking, one normally knows that one is walking, without considering the proposition. Knowing is a state, not an activity. In that sense, one can know without considering that things appear to be some way. (Williamson (2000), p. 198)

The view suggested in that passage is not Psychologism: the idea is not that the experience itself is evidence for its content, but rather that knowledge (albeit unconsidered knowledge) of the experience serves as an inferential base for belief in its content. The view is a version of Classical Foundationalism, according to which beliefs about one's experience are what justify beliefs about the external world. Williamson's complete view is therefore disjunctivist: Factualism about the good cases and Classical Foundationalism about the bad cases. There are different varieties of Classical Foundationalism: the Cartesian variety holds that beliefs about our own experiences are infallible, and that only deduction is allowed as a provider of inferential knowledge. A more modest variety lifts the infallibility claim and allows induction as well as deduction to aid in the acquisition of inferential justification. Williamson's variety seems to be in between these two: infallibilism about the basic beliefs (because they consist of known propositions), but with ampliative inferences allowed.

Classical Foundationalism has not been getting a lot of attention lately. One fundamental objection to it is that it seems to lead to a pretty severe form of skepticism. Suppose we grant that we do have rational beliefs, and even knowledge, about our own experiences—something that, Williamson's appeal to implicit knowledge notwithstanding, I am not so sure we should grant. Still, the vast majority of us (and all of us the vast majority of the time) do not form our beliefs about external objects on the basis of those beliefs about our experience of them. And if it is only beliefs about our experiences of external objects that justify beliefs about them, then most of our beliefs about them will not be *ex post* justified (they will not be based on the evidence which justifies them). Given that knowledge requires *ex post* justified beliefs, we know almost nothing about external objects. Williamson's combination of Factualism about the good case and Classical Foundationalism about the bad case is particularly puzzling. For, as I argued in chapter 4, this combination means that in order to be rational a subject would have to guess whether he is in the good case or the bad case, and then form his beliefs accordingly.

For these reasons, I will not be further discussing Classical Foundationalism, but rather concentrate on Psychologism, Factualism, and Experientialism.[7]

6.3 Against Psychologism

According to Psychologism, experiences themselves, not their contents, are basic reasons—they are our basic evidence when it comes to empirical justification. One fundamental role that evidence plays in the theory under development in this book is that of determining which states a subject can ignore in deciding what to do. For the states that the subject *cannot* ignore are those that are compatible with his evidence. If we go along with this function of evidence, Psychologism is not very attractive. For many states compatible with our experiences are ones that we do not normally consider when deciding what to do. We do not, for instance, consider the possibility that we are victims of an evil demon when making decisions. Now, it is open to the defender of Psychologism to say that we *should* consider that possibility when making decisions (as well as many other bizarre possibilities). But to take this route is to risk extensional inadequacy, or at least widespread skepticism about practical rationality.

Psychologism has it that we should consider many more states than we actually do in making our decisions. It is at least a live possibility that which action is rational when considering those states differs from which action is rational when not considering them. For instance, it may well be that considering the possibility that I am a brain in a vat changes whether it is rational for me to intend to eat this whole cake in front of me. After all, if I am a brain in a vat intending to eat the cake

[7] Byrne (2016) helpfully discusses similar views under the headings of the Traditional View (for my Classical Foundationalism), the Modern View (for my Psychologism), and the Postmodern View (for my Factualism). Experientialism doesn't seem to be on Byrne's radar.

will result in pleasurable culinary experiences, whereas if I am not the pleasurable experiences will be accompanied by sufficient damage to my health to outweigh those pleasurable experiences according to my utility function. Maybe the defender of Psychologism has a way of avoiding this kind of extensional inadequacy, but I find it hard to see how. Abstractly put, the problem is that more states have to be given positive credence according to Psychologism than according to either Experientialism or Factualism. It will then always be possible to construct a decision problem where this difference makes a difference with respect to the expected utility of the actions involved.

But even setting aside the problem of extensional adequacy, Psychologism is subject to an objection similar to the one I raised against Classical Foundationalism in the previous section. That objection was that Classical Foundationalism leads to skepticism because it has the consequence that few (if any) of our beliefs about external objects are *ex post* justified. Regardless of whether considering possibilities (such as my being a brain in a vat) leads to different results regarding the expected utility of the actions involved, normal people simply do not consider that possibility in their decision-making, just as normal people do not infer their beliefs about external objects from beliefs about their own experiences. And just as believing a proposition that is *ex ante* justified for a subject is not sufficient for that subject to be *ex post* justified, forming an intention to perform an action that maximizes expected utility is not sufficient for that intention to be *ex post* justified. There are delicate issues, both in the theoretical as well as in the practical case, about exactly what else is required for an *ex ante* rational attitude to be also an *ex post* rational attitude. But it seems eminently plausible that if the rationality of an intention is determined in part by the consequences that the action has if a state obtains, then forming the intention while ignoring that state will make it *ex post* irrational. Therefore, Psychologism has the consequence that almost every decision made by normal people turns out the be *ex post* irrational.

The second objection to Psychologism is simply that it is committed to implausible claims about evidential support. Notice that paradigmatic cases of evidential support are nothing like what Psychologism is committed to. There is a connection between evidential support and the intuitive notion of a "good argument." A set of propositions *P* constitutes evidence for a proposition *C* just in case the argument with the members of *P* as premises and *C* as its conclusion is a good one. Thus, for instance, paradigmatic cases of evidential support include cases of deductive inference such as instances of Modus Ponens with sufficiently simple premises and conclusion, or cases of inductive inference such as instances of enumerative induction with sufficiently simple premises and conclusion. For example, the propositions that Lucy is in her office and that if Lucy is in her office then the lights are on provide evidence for the proposition that the lights are on; and the propositions that a large and varied number of emeralds have been observed and they are all green provide evidence for the proposition that the next emerald to be observed will also be green. In each case, the corresponding argument is a good one. According to Psychologism, to these paradigmatic examples we have to add that propositions about subjects undergoing experiences are evidence for the propositions that are the contents of those experiences. For instance, Psychologism is committed to saying that the (true) proposition that Lucy has an experience with the content that the lights are on provides evidence for the proposition that the lights are on. But why would the fact that Lucy has an experience provide evidence for the proposition that is the content of that experience? The argument "Lucy has an experience with the content that the lights are on; therefore, the lights are on" is not good in the same sense in which the other two are. Of course, if we add to that argument a premise to the effect that Lucy's experiences are reliable then we do have a good argument. But that argument corresponds to what a version of Classical Foundationalism would say, not to Psychologism.

Why is the Psychologism-based argument not a good one? Perhaps the following consideration would help: the premises of that argument do not share a subject matter with its conclusion. This subject matter constraint is primarily a constraint on objective evidential relations. It holds that a set of propositions S supports a further proposition p only if, collectively, the propositions in S have the same subject matter as p. In the objective Bayesian framework, $Cr_u(p|S) > Cr_u(p)$ only if S and p share subject matter.[8] The subject matter constraint is imposed collectively on the set of propositions S, and for a good reason. It may be that a proposition that p does not have the same subject matter as a proposition that q, and yet it is part of a set of propositions S which do, collectively, have the same subject matter as q. Take, for instance, the propositions that Mary got her dream job and that John is happy. They have different subject matters: the first one is about Mary and her job; the second one is about John and his happiness. But if I add to the first one the proposition that Mary's getting her dream job would make John happy, then the result is a set of propositions that do share subject matter with the proposition that John is happy.

For the proponent of Psychologism, the proposition that the subject has an experience with the content that p is what justifies him in believing that p. But the subject matter of the proposition that the subject has an experience with the content that p is the subject himself and his experience, whereas p need not be about the subject at all (and, in particular, it need not be about the subject's experiences). Of course, the experience itself (I am assuming) does have a subject matter, its content, and the subject matter of the experience not coincidentally is the same as the subject matter of p. But to conclude from this that the proposition that the subject is having an experience with the content that p has the same

[8] For convenience, I adopt the convention that within the scope of a probability operator a set of sentences is to be understood as the conjunction of its members.

content as *p* itself would be a mistake. That the subject is having an experience with the content that *p* is no more a proposition about the fact that *p* than that the *New York Times* published an editorial about gun control is a proposition about gun control. Of course, the proposition that the *New York Times* published an editorial about gun control might be part of a set whose collective subject matter is indeed gun control. For instance, the set of propositions *The New York Times published an editorial about gun control* and *When the New York Times publishes an editorial about something, it becomes an important topic* are, collectively, about gun control. But, for the proponent of Psychologism, it is the fact that the subject has the experience that by itself justifies him in believing that *p*, not a set of facts of which the one about the experience is a member.

A third worry about Psychologism has to do with its claim that experiences provide reasons simply in virtue of existing—that is to say, it is sufficient for an experience to be a reason for a subject that the subject undergo the experience. Or, to put it in propositional terms, the proposition that the subject is undergoing an experience is a reason the subject has if and only if it is true. Compare this with the inferential case. In the inferential case, there is a clear distinction between the reason and the having of the reason—the proposition is the reason, and it is the subject's reason because he rationally believes it. If the proponent of Psychologism is right, then things are radically different in the non-inferential case: the reason and the having of the reason are one and the same thing. Maybe things could be said to mitigate the strangeness of this claim, but it is a *prima facie* additional cost of Psychologism.

Compare Psychologism with Experientialism in this respect. According to Experientialism, evidence consists of propositions, and to have some evidence a subject must be *ex ante* justified in believing the proposition in question. When the evidence is non-basic, it is based on some other evidence the subject has. For example, our reasons for thinking that all emeralds are green are propositions to the effect that many emeralds have been observed, and that they are all green. Our reasons are propositions, and we have them because we are justified in believing them. In the basic case, the evidence is not based on any further evidence. For instance, that there is a laptop in front of me is a proposition I am justified in believing, but not in virtue of being justified in believing some further propositions that are reasons for it. According to Psychologism, by contrast, this still doesn't count as a case of basic (i.e., non-evidential) justification, for I do have evidence that there is a laptop in front of me, and my evidence is that I have an experience with the content that there is a laptop in front of me.

Experientialism, then, has a unified account of evidence and its possession: evidence consists of propositions and it is possessed by a subject to the extent that he has basic *ex ante* justification for believing them. Psychologism, by contrast, has a disjunctivist account both of evidence and of its possession. Sometimes, evidence consists of propositions, sometimes of experiences. When evidence is propositional, to possess it is to be justified in believing it, whereas when evidence is experiential, to possess it is just to host it. To this extent, Experientialism is, if nothing else, more elegant than Psychologism. It also fits better with the connection between evidence and good arguments already alluded to. No matter how good an argument is, it will not justify you in believing its conclusion unless you are justified in believing is premises. Experientialism has no problem analogizing evidence to the premises of a good argument in that way. For Psychologism, however, the analogy has to be stretched to an uncomfortable extent. Sometimes, the premises of an argument are not propositions but experiences, and for the argument to justify you in believing its conclusion you need not believe its premises, but rather host them.

Some proponents of Psychologism, most prominently Pollock, evidently dislike this disjunctive account of evidence and its possession. Pollock unifies the account in what I take to be the wrong direction, however, by claiming that even in non-basic cases my reasons are not propositions but rather mental states. This comes at a cost. There is unification at one level, for evidence always consists of mental states, but the disjunctivism is still there just under the surface, for some of those mental states will be beliefs and other experiences. Moreover, as we shall soon see, Pollock needs to contort his theory in ugly ways when it comes to accounting for defeaters, contortions that are unnecessary if one treats evidence as propositional.[9]

Having said that, let me now acknowledge that there are indeed important similarities between Psychologism and Experientialism. Both Experientialism and Psychologism hold that there is an epistemically relevant relation between experiences and certain propositions.[10] Experientialism has it that experiences provide the subject with propositions about the external world as basic reasons. Psychologism has it that experiences are the basic reasons, and they are reasons for propositions about the external world. Does this mean that the difference between Psychologism and Experientialism is purely verbal? I doubt it. To some extent it is a matter of convention what it is that we call "evidence," but I have been explicit about the twin roles that the notion plays in *my* theory: the propositions that constitute our evidence are those that we are basically justified in believing, and they are also the propositions that determine which states we should consider when deciding what to do. It is these two roles, and not the word "evidence," that are of primary importance. If proponents of Psychologism agree with me on which propositions play these two roles, then more power to Psychologism.

[9] Turri (2009) argues for the view that reasons are mental states. In my view, his discussion is marred by a failure to take proper notice of the distinction between evidence and its possession.

[10] Factualists may (and perhaps should) also hold that there is an epistemically relevant relation between experiences and their contents.

6.4 Against Factualism

The troubles with Factualism can be more easily seen in the bad cases: cases where the content of the experience is false, but things are otherwise as in the good case. My arguments against Factualism parallel the arguments against knowledge-based decision theory given in the previous chapters.

Suppose that you want to drink a Crush, and you know that you can do so by pushing the button labeled "Crush"—but not, of course, the one labeled "Coke." Due to a bizarre interaction between your brain and a nearby radiation source, your experience represents the buttons as having their labels interchanged—the button labeled "Crush" is represented as being labeled "Coke" and vice versa. About this case, I want to say two things: first, it is rational for you to push the button labeled "Coke" (which your experience represents, remember, as being labeled "Crush"); second, given Fumerton's thesis, it is rational for you to push that button only if you rationally believe that it is labeled "Crush." Therefore, your belief that the button is labeled "Crush" is rational.

Factualists have to disagree. They have to disagree because your belief that the button is labeled "Crush" is false, and thus it doesn't amount to knowledge. So, Factualists must either deny that it is rational for you to push the button labeled "Coke," or they have to deny that rational action requires rational belief.

Suppose first that they take this second route, and deny Fumerton's thesis. A first concern is that this would be out of character for the Factualist. The natural ally of the Factualist is the knowledge-based decision theorist. But the knowledge-based decision theorist does believe that rational action requires rational belief, for he thinks that rational action requires knowledge, and does not deny that knowledge requires rational belief (indeed, as we have seen, many knowledge-first advocates simply identify rational belief with knowledge). But wave off that concern and suppose that the Factualist does indeed part ways with the knowledge-based decision theorist and claims that rational action does not require rational belief—in some cases, mere belief (whether rational or not) is enough. This is the view of Parfit, criticized in chapter 4. In that chapter, I noticed that one argument (the only one I could think of) in favor of the view is a bad one. For the only argument I could think of was that if I hold that rational action requires rational belief then I need to accept the existence of rational dilemmas—for, when the subject has irrational beliefs that bear on his actions, there is no action it is rational for him to perform. Recall my reply: true, the view does incur a commitment to rational dilemmas, but the kind of dilemmas to which the view is committed is not problematic, for they are merely *ex post*, self-imposed, and recoverable dilemmas. Therefore, there is no heavy price to pay for holding that rational action requires rational belief.

Suppose, then, that the Factualist takes the first route, and claims that it is not rational for you to push the button labeled "Coke." Maybe this action is excused, the Factualist could say, but it is not rational. But, as I argued when criticizing the excuses maneuver in chapter 5, whereas it may be all well and good for the Factualist to say that it is not rational for you to push that button, he needs to answer an urgent question: if that is not the rational thing for you to do in the situation, then what is the rational thing for you to do in the situation? As in the previous chapter, the options are limited and unappealing.

The first option is to say that it is rational for you to push the button labeled "Crush." But this is not an option that the Factualist himself can appeal to. For, just as you do not know that the button labeled "Coke" will not give you a Crush, because it won't, you also don't know that the button labeled "Crush" will give you a Crush, because your experience represents it as being labeled "Coke."

The second option is to say that it is rational for you to abstain from pushing either button. This option dovetails better with the Factualist view, for you do not know, of either button, that it will give you a Crush. But, again as in the previous chapter, we need to consider your epistemic position with respect to the proposition that the button labeled "Coke" (but which appears to you to be labeled "Crush") will give you a Crush. Of course, you do not know this proposition, because it is false. But, in your situation, it is not possible for you to know that you do not know this proposition. Even if you believed that you do not know it, this belief would not itself amount to knowledge, for it would be irrationally held. And notice that this is the case for everyone, not just for someone with my view. It would be absurd for anyone to claim that you know that you do not know the proposition in question. But then, the rationality of abstaining from pushing the button in question is determined by a condition that is unknowable to you—indeed, by a condition that it would be irrational for you to believe it obtains.

The final option is for the Factualist to say that there is no action or omission of yours that would be rational in this situation. As before, this amounts to positing the existence (and, indeed, abundance) of *ex ante*, world-imposed, unrecoverable rational dilemmas.

Unsurprisingly, then, Factualism runs into the same problems as knowledge-based decision theory does. Neither can satisfactorily answer the question about what is rational for a deceived agent to do.

Given the assumption that knowledge entails rational belief, Factualism and my own view, Experientialism, agree regarding good cases. But, as I argue in the next section, in the case of Factualism this extensional adequacy is fortuitous. Given that rational belief is sufficient for reason-possession in the bad case, it is sufficient for reason-possession in the good case as well.

6.5 For Experientialism

Experientialism holds (with Psychologism) that there is no rational difference between the good and the bad cases, but also that they are both cases of basic justification. In cases of basic perceptual justification, no evidence justifies the subject in either the good or the bad case. As I said before, the rational belief in question might well be the very first piece of evidence the subject receives. But what makes it the case that they have that piece of evidence in their possession is not their knowing the proposition in question, but rather their being justified in believing it.

Experientialism gets both the good and the bad case right. It gets the bad case right because it holds that what it is rational for you to do is to push the button labeled "Coke." As we saw in the previous section, that is the only viable option for what it is rational for you to do in this situation. It gets the good case right not just because it has the right consequence for that case (namely, that you should act on what you rationally believe), but also because it satisfies the subject matter constraint. What justifies you in pushing the button labeled "Crush" in the good case is not that you have an experience as of its being labeled "Crush" (this is a proposition about you), but rather the proposition that it is labeled "Crush," which is a proposition about the button itself.

Notice that the Experientialist would not say that the proposition that the button is labeled "Crush" is a reason the subject has for believing that the button is labeled "Crush." It is the proponent of Psychologism who thinks that the belief in question is evidentially acquired (as, according to him, are all empirically acquired beliefs). From the perspective of both Factualism and Experientialism, this is a mistake. The proposition that the button is labeled "Crush" is part of the subject's evidence, it is a reason he now has and on the basis of which he is now rational to believe further propositions and perform certain actions, but it is not itself evidentially supported. This doesn't mean, of course, that its being a reason that the subject has is unexplained. For the Factualist, the subject has this reason because he knows it, and for the Experientialist, the subject has this reason because he has an experience that justifies belief in its content, and this justification is not defeated. The proponent of Psychologism agrees with the Experientialist in that experience, and not knowledge, is what plays a crucial role in empirical justification, but the proponent of Psychologism thinks that the role is that of being evidence, whereas the Experientialist thinks that the role is that of providing evidence, of justifying without itself being evidence.

Now, in the good case I can in good conscience say that what justifies you in pushing the button is a fact about it, namely that it is labeled "Crush." In the bad case, however, there is no such fact. Still, you have the very same reason in the bad as in the good case: the proposition that the button is labeled "Crush." That reason is true in the good case, and false in the bad case.

In the next two sections I take up two objections to Experientialism: that it entails the possibility of inconsistent evidence, and that it cannot account for defeaters. Before moving on to those objections, I want to briefly consider a third one. This is the objection that, when I know that *r* is false, it sounds bad to say

"S's reason for ϕ $-ing$ was that r." To be honest, I am not very moved by that kind of objection. As I already said, the notion of a basic reason (or a bit of evidence) functions as somewhat of a technical term within Experientialism, defined mainly by its functional roles in the theory, the roles of being a proposition that a subject is justified in believing but not on the basis of being justified in believing other propositions, and the attendant role of determining which states should be considered when deciding what to do. It is not completely divorced from its meaning in ordinary language, of course, at least if Scanlon is a good guide to ordinary language when he says that the notion of a reason is that of a consideration that counts in favor (see Scanlon (2000)). But we should not expect it to agree with every nuance of ordinary use.

That said, there are perfectly kosher linguistic explanations for why reasons attributions appear to entail the truth of the reason attributed even when they do not.[11] Consider the following sentence:

(1) Barney's sister is tall.

Sentence (1) presupposes that Barney has a sister. Now, in general, contexts of belief-attribution function as presupposition plugs. Thus, if I say that Annie believes that Satan is evil, I have not normally presupposed that Satan exists. But sometimes presuppositions manage to leak through such contexts—or at least they appear to do so. Thus, consider:

(2) Barney believes that his sister is tall.

Sentence (2) appears to have two presuppositions: that Barney believes that he has a sister, and that Barney has a sister. Some linguists, however, have denied that (2) presupposes that Barney has a sister (see, for instance, Karttunen (1974) and Heim (1992)). Rather, what is going on according to these linguists is that (2) presupposes only that Barney believes that he has a sister, and from that information it can plausibly be inferred that he does indeed have a sister (that is not the kind of thing people are usually wrong about).

Something similar happens, the Experientialist can say, with attributions of reasons. Consider:

(3) Tomás's reason for tasting what Lucas was offering is that Lucas was offering candy.

Sentence (3) presupposes that Tomás believes that Lucas was offering candy. And, indeed, Experientialism backs this presupposition, for it holds that the proposition

[11] The following owes much to Matt McGrath—see Comesaña and McGrath (2014).

that Lucas is offering candy can be a reason Tomás has only if Tomás is justified in believing it. Now, it may seem as if (3) also presupposes that Lucas was offering candy. But we need not say that. We can instead say that all that is presupposed is that Tomás believes that Lucas was offering candy, and from that we can plausibly infer that Lucas was offering candy.

6.6 Inconsistent Evidence?

According to Experientialism, we can have false evidence. Doesn't that mean that, according to Experientialism, we can have inconsistent evidence? And isn't that a problem? Inconsistent evidence is incompatible, for instance, with traditional decision theory, for the states to consider when deciding what to do must partition logical space.

Williamson himself presented this objection when considering a view put forward by Goldman (2009). The objection is presented clearly and forcefully, and it is worth quoting in full (Williamson seems to be thinking about Escher's "Relativity"):

> On Goldman's account, it is not even clear that one's evidence must be *consistent*, let alone *true*. For some perceptual illusions involve unobvious inconsistencies in the perceptual appearances. For example, one may see a twisted closed loop of stairs S_0, S_1, S_2,. . ., S_n such that S_0 looks below S_1, S_1 looks below S_2,. . ., S_{n-1} looks below S_n and S_n looks below S_0. Since the stairs may not be especially salient, and there may be no special reason to trace this particular series round, spotting the inconsistency may take considerable ingenuity. Perhaps in these circumstances one is non-inferentially propositionally justified in believing that S_0 is below S_1, non-inferentially propositionally justified in believing that S_1 is below S_2,. . ., non-inferentially propositionally justified in believing that S_{n-1} is below S_n and non-inferentially propositionally justified in believing that S_n is below S_0 (even if one is not propositionally justified in believing the conjunction of those propositions). If so, on Goldman's equation E = NPJ, one's evidence contains the proposition that S_0 is below S_1, the proposition that S_1 is below S_2,. . ., the proposition that S_{n-1} is below S_n and the proposition that S_n is below S_0, in which case one's evidence is inconsistent, given that it is part of the logic of being below to be a transitive irreflexive relation. But there are grave difficulties in making sense of evidential probabilities on inconsistent evidence, since conditional probabilities are usually taken to be undefined when conditioned on something inconsistent. In particular, any proposition has probability 1 conditional on itself, and any contradiction has probability 0 conditional on anything (since conditional probabilities are probabilities): but these constraints cannot both be met for probabilities conditional on a contradiction (the conjunction of

> the inconsistent evidence propositions). Goldman does not say enough to enable us to tell how he would handle this case. Obviously it presents no difficulty to a view such as E = K on which all evidence is true, since then one's total evidence is automatically consistent. For a view on which not all evidence is true, the possibility of inconsistent evidence presents a significant threat.
>
> (Williamson (2009), p. 310)

Experientialism has it that one's evidence consists of those propositions which one has *undefeated* basic justification for believing. It is not obvious, in Williamson's example, that this condition is satisfied, for it is not obvious that the *prima facie* justification for believing each proposition is not defeated by the *prima facie* justification for believing each of the others.[12] But I myself gave the example of *knowingly* believing each of a set of inconsistent propositions in chapter 2 (or, if not believing, at least having probabilistically incoherent attitudes towards them). This was the case of my attitudes towards each of the three propositions involved in a particular case of the sorites: e.g., that Brian May is not bald; that one hair doesn't make the difference between being bald and not being bald; and that Yul Brynner was bald.

What, then, of the claim that "there are grave difficulties in making sense of evidential probabilities on inconsistent evidence"? My answer is that indeed there are such grave difficulties, and we better get used to it. If Williamson and I are right that it is possible to have undefeated basic justification to believe each of a set of inconsistent propositions, then we cannot make sense of evidential probabilities on what we are basically justified in believing—and that is enough of a problem even if you do not agree with me that your evidence is constituted by what you are basically justified in believing.

It is well worth remembering here that I do not officially take any form of Bayesianism to be correct, precisely because of the problem of logical omniscience. One manifestation of the problem of logical omniscience has to do with our computational limitations: we may be justified in believing inconsistent propositions because we just don't have the computational resources to spot the inconsistency. We can call this the omissive problem of logical omniscience. There is also, I have argued, a commissive problem: we may be justified in believing what we very well know to be an inconsistent set of propositions. For these reasons, no form of Bayesianism can be the literal truth. Nevertheless, I agreed with the many philosophers who take it to be a useful idealization. Useful for what? For figuring out the peculiar contribution that our ignorance of matters of fact makes

[12] It is interesting to note, in this respect, that Williamson leaves open the possibility that while the subject may be justified in believing each member of the inconsistent set, she may not be justified in believing their conjunction. But this seems to indicate that although the subject may have *prima facie* justification for believing each proposition, she doesn't have *ultima facie* justification.

to the rationality of our actions. When we are faced with either an omissive or commissive violation of logical omniscience, Bayesian decision theory cannot help us figure out what the rational actions are given our beliefs. Some may see this as an indictment of Bayesian decision theory; others, as an indictment of the view that there can be such violations of logical omniscience. I see it as neither. There are indeed such violations of logical omniscience, but (in fairness to Bayesian decision theory) it is not clear at all which actions (if any) those inconsistent beliefs rationalize. It may well be that, when our actions depend on our (basically justified) inconsistent beliefs, no such action can be properly said to be rational. Suppose, for instance, that something turns on whether someone who is a borderline case of baldness is in fact bald or not. It wouldn't be rational to act as if he is bald—nor would it be rational to act as if he is not bald. Such is life with justified inconsistent beliefs.

In sum, then: our having inconsistent evidence is compatible with Experientialism only if we can have undefeated basic justification for believing each of a set of inconsistent propositions. But if we can have undefeated basic justification for believing each of a set of inconsistent propositions, that is not a problem for the theory—or, at any rate, not any bigger problem than the one we already have if we accept that we can be basically justified in believing inconsistent propositions. And if we cannot be basically justified in believing inconsistent propositions, then the existence of inconsistent evidence is not compatible with Experientialism.

6.7 Defeaters

Another important objection to Experientialism is that it can't deal with the defeasibility of empirical justification.[13] The objection is important because coming to terms with it will force us to think carefully about defeaters and their role in epistemology.

The objection will take some time to develop. I start by delineating a very influential account of defeaters, that of John Pollock, who famously distinguished between rebutting and undercutting defeaters. Pollock's account, we shall see, is not without problems. There is some wavering regarding whether reasons are always mental states or (sometimes?) their content. More importantly, Pollock discussed only what I will call endogenous defeaters. Roughly speaking,

[13] I first heard of the objection from Stew Cohen in conversation, and then from Schroeder (forthcoming). A reader for OUP alerted me to the existence of Weisberg (2009) and Weisberg (2015), which reinforced my view that the objection affects not just Experientialism but Psychologism as well (a point that Weisberg makes), and, actually, pretty much any epistemological theory you have heard of (a point with which Weisberg might not agree). I also take from Weisberg the exogenous/endogenous terminology, which he applies to support rather than defeaters. My views on defeaters have changed since Comesaña (forthcoming), thanks in part to that anonymous reader.

endogenous defeaters work downstream from our evidence: they defeat the justification that our evidence gives us to believe (further) propositions. But, arguably, there are also exogenous defeaters. Roughly speaking, exogenous defeaters work upstream of our evidence: they defeat the power of our experience (or any other provider of evidence) to provide us with evidence. Of course, whether a defeater counts as endogenous or exogenous will depend on the theory of evidence under consideration—the same proposition that counts as an endogenous defeater for Psychologism, for instance, might count as an exogenous defeater for Experientialism. But any theory will have its own exogenous defeaters.

The objection to Experientialism is, in effect, that it cannot deal with exogenous undercutting defeaters. This I will grant. But the problem is more general: no theory of evidence couched in a Bayesian framework can deal with exogenous undercutting defeaters, whether it be Experientialism, Psychologism, or Factualism.[14] More than that, exogenous undercutting defeaters are a problem for a much more general group of theories: any theory according to which the justificatory power of evidence for a subject cannot depend on whether the subject has that evidence. Of course, whether a subject is *ex post* justified in believing something on the basis of some evidence does depend on whether she has that evidence, but the assumption in question is that the propositions that can be *ex post* justified by the evidence are precisely those that are *ex ante* justified by the same evidence.

6.7.1 Pollock on Reasons and Defeaters

Pollock thinks that all reasons are mental states—some are beliefs and some are experiences. The primitive in Pollock's system is the notion of *possible* ex post *justification*. More precisely, it is the notion that *it is logically possible for a subject S to be justified in believing a proposition P on the basis of mental state M*. In terms of this primitive, Pollock defines the notion of a reason:[15]

Reason: A state *M* of a person *S* is a *reason* for *S* to believe *Q* if and only if it is logically possible for *S* to become justified in believing *Q* by believing it on the basis of being in state *M*.

An understandable question at this point is: what does Pollock mean by saying that it is (logically) possible for *S* to be justified in believing *Q* on the basis of *M*? Does he mean on the basis of *M alone*, or on the basis of *M* together perhaps with

[14] This is, in effect, the point made by Pryor (2013a). But, in my terminology, the Bayesian-based objection to Pryor's dogmatism that he is answering in that paper (an objection first made by White (2006)) is that the theory cannot deal even with *endogenous* defeaters. More on this in chapter 8.

[15] I take the formulations from Pollock and Cruz (1999). Previous formulations can be found in Pollock (1970) and Pollock (1974).

some other mental states? The answer *has* to be "on the basis of *M* alone," because if we allow *M* to interact with other mental states, this overgenerates justification implausibly. For assume that *M* is a (perhaps conjunctive) justified belief. Is it logically possible for a subject *S* to be justified in believing an arbitrary *Q* on the basis of *M*, together with other mental states? Except perhaps for the cases where *Q* is incompatible with *M*, the answer will be "Yes." It will of course be logically possible for a subject to be justified in believing *If M, then Q*, and it is of course possible to be justified in believing *Q* on the basis of *M* plus this other mental state. So, Pollock must mean that for *M* to be a reason for *S* to believe *Q* it must be possible for *S* to be justified in believing *Q* on the basis of *M* alone. This atomism of Pollock's system is not particularly plausible, but I won't press this objection here.

There is another way in which Pollock's definition of a reason badly overgenerates. Even if one thinks that it is beliefs themselves that justify, rather than their contents, it should only be justified beliefs that justify. This is the principle of inferential justification that I've already alluded to in chapter 3.[16] But Pollock's official definition of a reason allows unjustified beliefs to be reasons. For of course it is logically possible for an unjustified belief to be justified. Take, then, an unjustified belief, and plug it in for *M* in Pollock's definition. For instance, to use an example from Pollock himself to which we will return, suppose that *M* is the belief that 87 percent of a random sample of voters from Indianapolis intend to vote Republican in the next election. On the basis of this belief, the subject concludes that 87 percent of the voters in Indianapolis will vote Republican in the next election. But suppose that the subject in question has the initial belief out of wishful thinking, and not on the basis of anything that remotely justifies it. It is still obviously logically possible for the subject to be justified in believing that 87 percent of the voters in Indianapolis will vote Republican in the next election, for it is logically possible for *M* to be justified for the subject (even if it actually isn't).

I foresee two ways for Pollock to solve this issue. One would be to hold that beliefs have their epistemic status essentially—i.e., that an unjustified belief is just a different state from a justified one. This move seems obviously *ad hoc* to me. Another solution is to give an explicitly recursive definition of a reason, where a mental state is a reason if and only if it is either an experience or it is justified by a reason (where justification is defined as having ultimately undefeated reasons). It is a delicate matter how exactly to formulate the recursive definition, but given that I think it is pretty clear that this last option is the one that better fits the spirit of Pollock's theory, I will proceed as if this is the view.

[16] But cf. Schroeder (2007), who argues that all beliefs justify, but unjustified beliefs are guaranteed to have their justificatory power defeated.

Pollock then defines the general notion of a defeater:

Defeater: If M is a reason for S to believe Q, a state M^* is a *defeater* for this reason if and only if the combined state of being in both the state M and the state M^* at the same time is not a reason for S to believe Q.

Unpacking the definition of reason gives us the following:

Defeater, unpacked: If it is logically possible for S to become justified in believing Q by believing it on the basis of being in state M, then state M^* is a *defeater* for this reason if and only if it is not logically possible for S to become justified in believing Q by believing it on the basis of being in the combined state M *and* M^*.

The definition is stated in a conditional form, but what Pollock is really doing here is defining the notion of *M^*'s being a defeater for M as a reason for S to believe Q.* For that to happen, M has to be a reason for S to believe Q, and the combination of M with the defeater state not be a reason for S to believe Q. The fact that what Pollock is defining is this four-place relation will become important to our discussion.

Famously, Pollock distinguished two kinds of defeaters. One of them, he says, appealing to the analogy with arguments we used before, "is a reason for denying the conclusion" (Pollock and Cruz (1999), p. 196):

Rebutting defeater: If M is a reason for S to believe Q, M^* is a *rebutting defeater* for this reason if and only if M^* is a defeater (for M as a reason for S to believe Q) and M^* is a reason for S to believe $\neg Q$.

Thus, *believing that Lucy is in her office and that most of the times when she is in her office Lucy turns on the lights (P)* is a reason for believing that *the lights in Lucy's office are on (Q)*, but believing that *Joe was just outside Lucy's office and he reports seeing through the window that the lights were off (D)* is a rebutting defeater for our belief in P as a reason to believe Q. For we cannot become justified in believing C by believing it on the basis of our beliefs that P and that D, and belief in D is by itself a reason to believe $\neg Q$.

The second kind of defeater discussed by Pollock "attacks the connection between the premises and the conclusion rather than the conclusion itself" (Pollock and Cruz (1999), p. 196). Here we must pause to notice the awkwardness in Pollock's theory which results from treating all reasons as mental states. The example he gives of a rebutting defeater is one we already mentioned—that of a pollster attempting to predict what proportion of residents of Indianapolis will vote for the Republican candidate in the upcoming election:

> She randomly selects a sample of voters and determines that 87% of those polled intend to vote Republican. *This* gives her a defeasible reason for thinking that approximately 87% of all Indianapolis voters will vote Republican. But then it is discovered that purely by chance, her randomly chosen sample turned out to consist exclusively of voters with incomes of over $100,000 a year. *This* constitutes a defeater for the inductive reasoning...
>
> (Pollock and Cruz (1999), p. 196, my emphases)

Notice the unclarity about the referent of the two emphasized appearances of "this." What is the initial reason, and what is the defeater? The official story must be that they are mental states of the pollster, but the natural way to read the sentences is as referring to the *propositions* that 87 percent of those polled will vote Republican and that all those polled have incomes of over $100,000 a year. And it is not just the natural interpretation of those sentences. After the passage quoted Pollock gives the following definition:

Undercutting defeater: If believing P is a defeasible reason for S to believe Q, M^* is an *undercutting defeater* for this reason if and only if M^* is a defeater (for believing P as a reason for S to believe Q) and M^* is a reason for S to doubt or deny that P would not be true unless Q were true.

There is a lot to go through in that definition. First, notice that Pollock sticks to the official line that reasons are mental states when it comes to the initial reason and the defeater. The initial reason is *believing that P*, and the defeater is some unspecified mental state M^*. Instantiating the definition to Pollock's own case, the initial reason is the pollster's belief that 87 percent of those polled will vote Republican, and M^* is presumably the pollster's belief that all of those polled have incomes of over $100,000 a year. Notice also that the choice for the initial reason is a poor one. As I already noted, presumably an *unjustified* belief that 87 percent of those polled will vote Republican is not a reason to believe that 87 percent of the relevant population will vote Republican—but bracket that issue. Another remarkable feature of the definition is that according to it undercutting defeaters are reasons to doubt or deny a counterfactual conditional. The obvious way of defining an undercutting defeater (setting aside for a moment the problem to be presented in the next paragraph) would have been, instead, the following: M^* is an undercutting defeater for M as a reason for S to believe Q if and only if M^* is a defeater (for M as a reason for S to believe Q) and M^* is a reason for S to think that M is not a reason for S to believe Q. Presumably, Pollock appealed to the counterfactual instead of to this explicitly epistemic definition because he thought the latter definition "too intellectualistic," in that it would only apply to subjects who themselves have the notion of a reason.

The most remarkable feature of the definition of an undercutting defeater, however, is that it just does not fit the informal gloss of the notion given by

Pollock himself before the definition. To fit the informal characterization of an undercutting defeater as "attacking the connection between premises and conclusion," M^* would have to be a reason for thinking that the premises would not be true unless the conclusion were true. Both the premises and the conclusion, sticking to the official story, are beliefs of the pollster. Therefore, the last clause of the definition would have to be "M^* is a reason for S to doubt or deny that S would believe P unless S believed Q." But, of course, that clause wouldn't make sense. What S would or wouldn't believe under different conditions is irrelevant to whether he has a defeater. Hence the weirdly disjointed definition, which sticks to the official story for the original reason and the defeater, but defects to thinking of reasons as the propositions believed, rather than the believing of them, in the last clause.

Why would Pollock do this? A clue comes from the fact that immediately after giving that definition Pollock says: "When the reason is a nondoxastic state, we must define undercutting defeat slightly differently," and gives this other definition:

Nondoxastic undercutting defeat: If M is a nondoxastic state that is a defeasible reason for S to believe Q, M^* is an *undercutting* defeater for this reason if and only if M^* is a defeater (for M as a reason for S to believe Q) and M^* is a reason for S to doubt or deny that he or she would not be in state M unless Q were true.

This is the real reason Pollock takes reasons to be states rather than their contents, and the reason for all the awkwardness in his definitions I have been pointing to: to take care of the case of justification by experience and its defeat. This difference with the doxastic case deserves extended discussion rather than what amounts to barely an acknowledgment that there is a difference. For Experientialism, of course, there is no such difference, for evidence is propositional through and through.

Let me now turn to the question of how to capture defeaters in a Bayesian framework. The disappointing answer will be that some cases of defeaters cannot be captured in that framework. But that will not turn out to be a problem for Bayesianism in particular, for the kinds of defeaters that cannot be captured in the Bayesian framework cannot be adequately treated in any other framework either, Pollock's own included.

6.7.2 Defeaters for Bayesianism

Let us start with a relatively easy case. Suppose that we have some evidence E for a proposition H, and that D is either a rebutting or an undercutting defeater for E as a reason for H. As anticipated, I will call defeaters of this kind *endogenous* defeaters. Can endogenous defeaters be captured in a Bayesian framework? Yes, they can. Recall that in the Objective Bayesian framework epistemic relations are captured in the ur-prior. Thus, the claim that E is evidence for H is captured in the claim that $Cr_u(H|E) > Cr_u(H)$, and the claim that D is a defeater for E as a reason for H is captured by adding to the previous constraint on the ur-prior the following one: $Cr_u(H|E \wedge D) < Cr_u(H|E)$. Let us check that these translations fit the standard cases of endogenous defeaters. First, consider an endogenous rebutting defeater. Let E be the conjunctive proposition that Lucy is in her office and that most times when Lucy is in her office the lights are on, H be the proposition that the lights in Lucy's office are on, and D the proposition that Joe reports that the lights in Lucy's office are off. The translations check out: our conditional credence that the lights in Lucy's office are on given that she is in it and that most of the time that she is in it the lights are on should be higher than our unconditional credence that the lights are on, but our credence that the lights are on conditional on that very same proposition *and* the proposition that Joe reports that the lights are off should be lower than the previous conditional credence—how much lower will depend on background knowledge regarding, for instance, Joe's reliability. Consider next Pollock's example of an undercutting defeater: let E be the proposition that 87 percent of those polled will vote Republican, H the proposition that 87 percent of the relevant population will vote Republican, and D the proposition that all of those polled have an income greater than $100,000. Again, the translation checks out: our conditional credence that 87 percent of the relevant population will vote Republican given that 87 percent of those polled will vote Republican should indeed be higher than our corresponding unconditional credence, but it should go down when we add the condition that all of those polled have an income greater than $100,000. There is still some work to be done to have a full account of endogenous defeaters within a Bayesian framework. There are interesting questions, for instance, about how to distinguish between undercutting and rebutting defeaters and how to incorporate other kinds of defeaters that Pollock might have missed. For a careful and revealing analysis of such issues, I refer the interested reader to Kotzen (forthcoming). But although such interesting questions of detail do remain, there don't seem to be insurmountable problems with capturing endogenous defeaters in a Bayesian framework.

Things are different with what I will call *exogenous* defeaters. The notion of an exogenous defeater is best introduced with an example. Suppose that I look at the wall and see that it is red, and on that basis believe that the wall is red. According to Experientialism (and Factualism) I now have as evidence the proposition that the wall is red.[17] Now suppose that a trusted friend who is right next to me tells me that the wall isn't red.[18] Alternatively, suppose that what the trusted friend tells me

[17] As I argue later, the same problem that I am about to pose for Experientialism and Factualism also arises for Psychologism, but the example will have to be different.

[18] For it to be true that I did see that the wall is red my friend has to be wrong, but all that matters is that I can be justified in trusting her.

is that the wall is bathed in red lights. What my friend tells me in the first case resembles a rebutting defeater, and what she tells me in the second case resembles an undercutting defeater. And yet, the propositions that the wall isn't red and that the wall is bathed in red lights (or that my friend tells me as much) do not fit the definition of defeater given by Pollock. For recall that, according to that definition, defeat is a four-place relation: D is a defeater for E as a reason for a subject S to believe H. In this case, however, we are missing one of the relata: we have the subject (myself), we have E (that the wall is red), and we have D (say, that the wall is bathed in red lights or that my friend says that the wall isn't red), but we are missing H. Of course, we could concoct an H if we wanted to. We can, for instance, let H be the proposition that the wall is crimson. But simply finding an H like that will not magically transform D into the right kind of defeater to fit Pollock's definition. Notice that the fact that there are red lights shinning on the walls is not a reason to doubt that the wall's being red is a reason to think that it is crimson. Even the claim that the wall is not red is not a reason to doubt that the proposition that it is red epistemically supports (is evidence for) the proposition that it is crimson. Rather, what D is doing in this case is lowering rational credence in E itself, not in E as a reason for believing anything else. This is why I call them exogenous defeaters: they defeat rational credence in our evidence itself, not in the reason-giving power of our evidence.

How could we incorporate exogenous defeaters into the Bayesian framework? Let us start with exogenous rebutting defeaters. Let E be the proposition that the wall is red and D the proposition that my friend tells me that the wall is not red, and let us assume that, initially, my credence that the wall is red is low (for all I know, say, it could be any one of thirty different colors). After I see the wall, E is now part of my evidence, and so my credence in E should be 1. But it should also be the case that I lower my credence in E after my friend tells me that the wall isn't red. Now, we know that if $Cr(E) = 1$, then $Cr(E|D) = 1$ also, and so it looks as if it will be impossible to capture this case in the Bayesian framework.

Here is where some people will be tempted to switch from the standard Bayesian account to one based on Jeffrey's idea that our evidence is not always propositional, and his corresponding generalization of Conditionalization (see Jeffrey (1965)). According to Jeffrey, our experience doesn't always rationalize our assigning credence 1 to some proposition, even if it does induce a change in our credence distribution. Thus, Jeffrey might say, after looking at the wall I shouldn't assign credence 1 to the proposition that the wall is red, but rather raise my credence in that proposition perhaps significantly, although leaving open the possibility that it might be some other color (say, one of those greens that I find difficult to distinguish from some reds). But if after undergoing that experience I do not assign credence 1 to any proposition, how should I propagate what I learned to the rest of my credences? I cannot leave everything else as it is, because that may well be a violation of Probabilism (and, more importantly, will not be giving my evidence its due), but I cannot conditionalize either, because there is nothing to conditionalize on. Jeffrey's reply is that when, as a result of my experience, my credences change from time i to time j only over a partition $\{E_1 \ldots E_n\}$, then instead of updating the rest of my credences by Conditionalization I should update them according to the following rule:

Jeffrey's Conditionalization: $Cr_j(A)\sum_{i=1}^{n} Cr_i(A|E_i)Cr_j(E_i)$.

Adopting this version of Bayesianism does allow us to capture exogenous rebutting defeaters. If after seeing the wall my credence that it is red should not go all the way up to 1, then there is no obstacle for it come back down again after hearing my friend say that it is not red. But I do not advocate a switch to a version of Bayesianism that incorporates Jeffrey's suggestions, for two reasons. First, as we shall soon see, it will not help with the case of exogenous undercutting defeaters. Second, an ur-prior-based Bayesianism already has the resources to incorporate exogenous defeaters

Let us then explore the remaining kind of defeater: exogenous undercutting defeaters. These turn out to be the most problematic kind. I begin by explaining a result by Weisberg (2015) which shows that exogenous undercutting defeaters cannot be incorporated into either the standard Bayesian framework or Jeffrey's version.

Both Conditionalization as well as Jeffrey Conditionalization have a property that Weisberg calls "rigidity": namely, that the conditional probabilities on the partition that generates the update are the same before and after the update. For Conditionalization, the partition that generates the update is simply $\{E, \neg E\}$, for some E, and the update always results in assigning 1 to E and 0 to $\neg E$. For Jeffrey Conditionalization, the partition might be more complex, and it need not result in the subject's assigning credence 1 to any proposition in the partition. For generality's sake, let us suppose that the update starts by changing credences on a partition of propositions $\{E_1 \ldots E_n\}$. Rigidity then guarantees that, for any proposition P, $Cr_a(P|E_i) = Cr_b(P|E_i)$ for each E_i, where Cr_b is the subject's credence function before the update and Cr_a is the subject's credence function after the update. Recall now the example of an exogenous undercutting defeater that I gave above. In this case, E is the proposition that the wall is red, and D is the proposition that there are red lights shining on the wall. Now, before looking at the wall, my credence that the wall is red is independent of my credence that it is illuminated by red lights, i.e., $Cr_b(E|D) = Cr_b(E)$. We want it to be the case that after I look at the wall these credences are no longer independent: $Cr_a(E|D) < Cr_a(E)$. Of course, according to Conditionalization this cannot happen, for $Cr_a(E) = 1$, and so $Cr_a(E|D) = 1$. But going with Jeffrey Conditionalization won't help this time. For suppose that we Jeffrey conditionalize on $\{E_1 \ldots E_n\}$ (one of the E_i will be the proposition that the wall is red, and its probability will presumably go up but not all the way to 1). We know that the color of the wall is independent of the lighting conditions, and so:

$$Cr_b(E_i|D) = Cr_b(E_i), \text{for each } E_i. \tag{1}$$

As before, we want it to be the case that after I look at the wall these are no longer independent: $Cr_a(E_i|D) < Cr_a(E_i)$, for at least some E_i. But, given rigidity:

$$Cr_a(D|E_i) = Cr_b(D|E_i), \text{for each } E_i. \tag{2}$$

Now, given (1) and the symmetry of independence:

$$Cr_b(D|E_i) = Cr_b(D), \text{for each } E_i. \tag{3}$$

Combining (2) and (3):

$$Cr_a(D|E_i) = Cr_b(D), \text{for each } E_i. \tag{4}$$

By the theorem of total probability:

$$Cr_a(D) = Cr_a(D|E_i)Cr_a(E_i), \text{for each } E_i. \tag{5}$$

By (4), however, each one of the $Cr_a(D|E_i)$ is the same value (namely, $Cr_b(D)$), and so the weighted average in (5) becomes:

$$Cr_a(D) = Cr_a(D|E_i), \text{for each } E_i. \tag{6}$$

And, given again the symmetry of independence:

$$Cr_a(E_i|D) = Cr_a(E_i), \text{for each } E_i. \tag{7}$$

As Weisberg puts it:

> Intuitively, of course, the trickiness of the lighting should be irrelevant to the [wall's] color before the glance, and negatively relevant after. But, as long as the update in between is rigid with respect to the [wall's] redness, this cannot happen. Rigidity prevents the introduction of a negative correlation between E and [D].
>
> (Weisberg (2015), p. 126)

Now, Psychologists might complain that the problem arises only because we have misidentified the evidence. If Psychologism is right, then the evidence is not that the wall is red, but something like the fact that the subject has an experience with the content that the wall is red. But Psychologism will not help with the problem of exogenous undercutting defeaters—it will at best move it one step backwards. For suppose that we agree with Psychologism that my evidence is that I'm having an experience with the content that the wall is red. There are exogenous undercutting defeaters for that proposition too. Suppose, for instance, that my friend tells me that my coffee has been spiked with a drug which causes me to confuse which experiences I'm currently having: people under the influence of that drug are often certain that they have an experience with the content that the object in front of them is red when in fact they have a completely different experience. If we let *E* now be the proposition that I am having an experience with the content that the wall is red and let *D* be the information about the drug, then the reasoning above goes through as before. In particular, before having the experience I should treat the information about the drug as irrelevant to my having the experience (the drug works by making me confuse one kind of experience with another, not by creating different experiences), but after having the experience I should treat it as defeating information. Rigidity, however, prevents this from happening.

Now, the proponent of Psychologism may well point out that, according to that view, the evidence is the experience itself, or perhaps a proposition to the effect that the subject is having the experience, but certainly not a *belief* by the subject that she is having the experience. The subject may have no beliefs whatsoever about her experiences, as long as she has the experiences themselves, and still have evidence. What is the bearing, then, of the information about the drug on a subject who does not have beliefs about her experiences?

That is a good question, but (a) if it helps Psychologism out of the problem of exogenous undercutting defeaters it does so by suggesting that there are no exogenous undercutting defeaters, and so it helps any other view as well, and (b) it does not actually help Psychologism. Sturgeon (2014) has argued that there are no undercutting defeaters as Pollock conceived of them. The complaint is that, at least when the subject has no beliefs about the origin of their experiences, what Pollock would count as undercutting defeaters are not defeaters after all. I take Sturgeon's argument to have at least some force when it comes to non-doxastic undercutting defeaters,[19] and although Sturgeon is not concerned with Experientialism, his point applies there as well. Suppose that I have no beliefs whatsoever about the etiology of my belief that the wall is red—I don't even have beliefs about what experiences I am having. Why would the information about the lighting then defeat my belief that the wall is red?

But Sturgeon's argument doesn't establish that there aren't any undercutting defeaters. The case where I have no beliefs whatsoever about the origin of my beliefs may not, after all, be the normal case. Perhaps in the normal case, I do believe (however implicitly and roughly) that I am looking at the wall, and that my belief that the wall is red is based on my looking at it. And if I do have those beliefs, then information about the lighting will obviously be relevant to the justification of my belief. The same goes for Psychologism. Perhaps normally we do not have beliefs about our experiences or how reliable we are about

[19] Although conversations with Matthew McGrath have made me doubt the relevance of Sturgeon's case of the "Milk taster" in Sturgeon (2014).

detecting them. But information about our reliability, like the information about the drug, does make us aware of the fact that we can be wrong about whether we have the relevant experiences. In either case, the reasoning above still goes through. I noticed before that whether I have an experience as of the wall's being red is independent of whether I take the drug or not. Similarly, whether I have an experience as of the wall's being red is also independent of whether I am a reliable detector of my own experiences. And yet, after having the experience I should take the combined information about the drug and my consequent unreliability as defeating my experience as evidence.[20]

I conclude that Psychologism suffers from the same problem as Factualism and Experientialism when it comes to incorporating exogenous undercutting defeaters. Indeed, the problem with exogenous undercutting defeaters runs deep, because it arises out of the difference between evidence and what it is evidence for, on the one hand, and evidence and the conditions under which it is had as evidence, on the other. Any theory will have to draw this distinction somewhere, and so every theory will have to deal with exogenous defeaters. One could try to incorporate providers of evidence within the scope of epistemic principles, and some work in that direction has already been done (see, for example Schwarz (2018) and Gallow (2013)), but it's fair to say that the jury is still out on the success of that strategy.

Let me now explain how appealing to ur-prior Conditionalization, instead of standard or Jeffrey Conditionalization, can help with the problem of exogenous defeaters. Let us suppose that I know that the color of the wall (red or white) is going to be decided by the flip of a fair coin, and the color of the lights (also red or white) is also going to be decided by a second flip of the same fair coin. If we let K represent all this background information, then, conditional on K, the ur-prior assigns the same probability (namely, ¼) to each of the state-descriptions determined by the partition $\{E, \neg E, D, \neg D\}$, where E is the proposition that the wall is red and D is the proposition that the lights are red.[21] So, in particular, $Cr_u(D|E) = Cr_u(D) = Cr_u(E|D) = Cr_u(E) = ½$. When I open my eyes and look at the wall, I have an experience with the content that E and, as a result, I have E as evidence. Therefore, letting Cr_1 be my credence function at this time, $Cr_1(E|D) = Cr_1(E)$. This might seem as the wrong result, for we want my credence that the wall is red to be affected by learning that the lights are red. But my credence will be affected when I later learn that the lights are red, for, letting Cr_2 be my credence function at that time, then $Cr_2(E) = Cr_u(E|D) < Cr_1(E|D)$. Thus,

[20] Proponents of Psychologism may also claim that there cannot be exogenous undercutting defeaters because our evidence is indefeasible. This, however, would be a return to a kind of Cartesianism which not many would be happy with.

[21] For convenience, in what follows I drop consideration of K.

appealing to ur-prior Conditionalization does allow us to introduce a dependence between D and *E*.

We can, therefore, incorporate rebutting defeaters in a Bayesian framework by opting for an ur-prior-based version of Experientialist Bayesianism. This way of dealing with exogenous defeaters highlights an important issue, however. That important issue is related to a well-known worry about Jeffrey Conditionalization (and, therefore, standard Conditionalization as well): that it treats the impact of experience on credences as an exogenous input to the Bayesian machinery. This is unfortunate to the extent that we think that there are substantive epistemological principles dealing with the experience-credence interface that our theory should capture. As for my own preferred view, it suffers from a similar problem. What is treated as an exogenous input is in this case the relationship between exogenous defeaters and evidence. We are simply told that I no longer have as evidence the proposition that the wall is red when my friend tells me about the red lighting, but we are not given an explanation as to why and how this is so. But notice that this issue would arise for Factualism and Psychologism as well, were they to deal with the problem of exogenous defeaters by appealing to ur-prior Conditionalization. And even setting aside Bayesianism, Pollock's own theory of defeaters treats the fact that *D* defeats *E* as an exogenous input to his theory, for we are simply told that it is not possible to be justified in believing *E* when in the combined mental state of having an experience with the content *E* and a belief with the content *D*. The problem of exogenous defeaters simply highlights what we already knew: that we cannot say anything more substantive as to why this is so.

6.8 Conclusion

In this chapter I have presented Experientialism as a theory of basic evidence. I argued that Experientialism fares better than both Factualism and Psychologism, for both of these latter theories would have us countenance too many states in deciding what to do. I have also examined whether defeaters pose a problem for Experientialism. One kind of defeaters, exogenous defeaters, do pose a problem for any theory wishing to incorporate them into a formal framework such as Bayesianism. But it is indeed a problem for any such theory, and so not a mark against Experientialism in particular. Moreover, it is not clear that a theory of justification should try to incorporate these kinds of defeaters into a formal framework, as opposed to treating them as an exogenous input to the formal machinery.

In chapter 5 I defended Fumerton's thesis against the view that unjustified beliefs can justify further beliefs and actions. Denying that claim, however, comes at a cost. For suppose that I irrationally believe that it is raining and, when confronted with the question whether it is precipitating, I either suspend judgment or disbelieve this other proposition. Even if my belief that it is raining is unjustified, it might seem that being incoherent and disbelieving that (or even suspending judgment on whether) it is precipitating is a further irrational attitude on my part. But why would it be, if unjustified attitudes cannot themselves justify? I take up that question in the next chapter.

7
The Normative Force of Unjustified Beliefs

7.1 Introduction

Many of the arguments presented throughout the last three chapters rested on Fumerton's thesis: the claim that a belief can justify an action, or intention, only if it is itself justified. I also alluded to the analogous thesis for the case of theoretical justification, which has sometimes been called the "principle of inferential justification": that a belief can justify another belief only if it is itself justified. Not everyone accepts those theses, however. Parfit, for instance, rejects Fumerton's thesis—although he accepts the principle of inferential justification. I argued that, in the Bayesian framework, one key role for the notion of belief is that of evidence. Many subjective Bayesians require so little of a proposition for it to be evidence for a subject (in the extreme, just that it be believed and that it not be a formal contradiction) that it is clear that they reject at least the spirit of both Fumerton's thesis and the principle of inferential justification.

John Broome has recast this dispute about Fumerton's thesis and the principle of inferential justification in different terms (see, for instance, Broome (1999)). To put it very roughly, Broome's notion of a normative requirement is, in effect, the notion of an attitude that would justify another attitude if it itself were justified.[1] The most straightforward interpretation of normative requirements raises some puzzles. Broome's project has at least two aspects: to give an interpretation of normative requirements that answers those puzzles, and to argue that they do carry some kind of normative force, even when the input attitudes are not themselves justified.[2]

Let me start with an example stated in the objective Bayesian framework. Suppose that a subject unjustifiedly believes that a representative sample of emeralds has been observed, and they have all been found to be blue. Call this proposition P, and call the proposition that the next emerald to be observed will also be blue, Q. Arguably, $Cr_u(Q|P) > Cr_u(Q)$. Assume also that the subject has no

[1] This characterization is only a very rough one because (a) Broome intends normative requirements to apply to actions as well as attitudes, and (b) Broome (2013) rejects normative detachment.

[2] It is not in fact obvious to me that Broome is engaged in this second aspect of the project. However, he has been interpreted as being so engaged by many, and I follow here that common interpretation.

Being Rational and Being Right. Juan Comesaña, Oxford University Press (2020). © Juan Comesaña.
DOI: 10.1093/oso/9780198847717.001.0001

(other?) evidence relevant to Q. And now consider two possible variations of the case. In one, the subject's credence in Q is $Cr_u(Q|P)$, whereas in the other is some different credence (for concreteness, suppose that it is $Cr_u(Q)$). It is easy to feel a certain tension about these variations. On the one hand, we want to say that, just as the subject is unjustified in believing P in the first place, he is also unjustified in assigning credence $Cr_u(Q|P)$ to Q. On the other hand, given that he does in fact believe P, he seems to only be adding insult to injury if he doesn't assign credence $Cr_u(Q|P)$ to Q.

Broome himself has given other examples. One mirrors the previous case in a binary framework. Thus, suppose that P obviously entails Q. Consider now the following principle:

Closure: If you believe that P, then you ought to believe that Q.

And here are other examples from Broome:

> If you believe that you ought to ϕ, then you ought to (intend to) ϕ. If you intend to ϕ, and believe that in order to ϕ you must ψ, then you ought to (intend to) ψ.
>
> If you believe that there is sufficient evidence for P, then you ought to believe that P.
>
> If you believe that P, then you ought not to believe that $not-P$.

On the most straightforward interpretation, normative requirements are conditionals. We will soon see that there are reasons to doubt this straightforward interpretation, but let us stick with it for a moment in order to introduce one of the puzzles to be discussed.

If normative requirements are conditionals, then their antecedents are about beliefs or intentions—for instance, the proposition that you believe that you ought to ϕ, or the proposition that you intend to ϕ. Their consequents, however, are all deontically modified propositions—propositions to the effect that you ought to intend to do something or you ought to believe some proposition. For each normative requirement, there is a parallel principle which is just like it except that its antecedent is deontically modified too. For instance, if we modify the antecedent of Closure in that way, the result is a version of the principle of Epistemic Closure that is widely discussed in the literature:[3]

Epistemic Closure: If you ought to believe that P, then you ought to believe that Q.

[3] The version of Epistemic Closure discussed in the epistemology literature is formulated most often with respect to knowledge or justification. How close the version I am about to formulate is to that one will depend, among other things, on how closely related one thinks "ought" statements are to justification. I discuss closure principles further in chapter 8.

To illustrate the difference between Closure and Epistemic Closure, suppose that we both believe that it is raining on a certain occasion, but whereas you believe it because you can see that it is raining, I believe it out of a gloomy disposition (I haven't yet opened my eyes this morning, say). Now consider the proposition that it is precipitating, which (I will take it) is obviously entailed by the proposition that it is raining. Whereas Epistemic Closure applies only to you, Closure applies to both of us, and so has as a consequence that our belief that it is raining requires us to believe that it is precipitating. But although this consequence might well be the correct one in your case, I should arguably recognize that I ought not to believe that it is raining and so abandon that belief, instead of adding insult to injury and believing in addition that it is precipitating. And yet, we do feel the pressure to say that if, contrary to reason, I insist on believing that it is raining, then at least I also ought to believe the obvious consequence of that proposition, that it is precipitating. A particularly vivid way of bringing out this pressure is to consider two subjects who, just like me, believe that it is raining when they have no business doing so. One of them, Mary, believes in addition that it is precipitating, whereas the other one, John, does not. Whereas we would fault both Mary and John for believing that it is raining, we cut Mary some slack for at least being consistent and also believing that it is precipitating. Analogous points can be made about the other examples of normative requirements.

That is the tension that I want to explore in this chapter: on the one hand, honoring Fumerton's thesis and the principle of inferential justification, we think that normative requirements cannot do any epistemic heavy lifting of their own (whether I ought to believe that it is precipitating depends on whether I ought to believe that it is raining, not merely on whether I do in fact believe it); on the other hand, however, we think that they do have some epistemic import (if you are going to believe that it is raining despite the evidence, then at least you ought to believe that it is precipitating). In this chapter I defend two main theses. First, the problem of interpreting normative requirements so as to explain this tension is intimately related to an older problem in deontic logic, the problem of contrary-to-duty (ctd) obligations. Second, the solution to both problems involves thinking of conditionals in a way that will be surprising to philosophers, but not at all to linguists.

I begin by clarifying what the problem of interpreting normative requirements consists in. To do so, I take a look at what Broome, who introduced the terminology of "normative requirements," has to say about them.

7.2 Broome's Interpretation of Normative Requirements

As I said, perhaps the most straightforward interpretation of Closure is as a conditional with a deontically modified consequent. So, where "B" is a belief operator to be interpreted as "believes that..." and "O" a deontic operator to be interpreted as "ought to make it the case that...," that interpretation of Closure can be rendered thusly:

Closure-N: $B(p) \rightarrow O(B(q))$.

I call it "Closure-N" because the deontic operator takes narrow scope over the consequent of the conditional. Broome has argued against narrow-scope interpretations of normative requirements in general. Suppose, first, that the conditional in question in Closure-N obeys Modus Ponens. This means that, according to Closure-N, simply believing (no matter for what reason) that it is raining entails that you ought to believe that it is precipitating. But this leads to nasty bootstrapping problems. For suppose that you ought not believe that it is raining but you (irrationally) believe it nonetheless. In that case, Closure-N entails that you ought to believe that it is raining (given that p obviously entails p). Even if you ought not to believe a proposition, the mere fact that you do believe it entails that you ought to do it. This is an absurd form of bootstrapping. If the conditional in question is the material conditional, there are additional problems for interpreting Closure as Closure-N. For suppose that you do not believe that it is raining. That entails, according to this interpretation of Closure, that believing that it is raining requires you to believe any arbitrary proposition (a material conditional is entailed by the negation of its antecedent).

In response, Broome argued that normative requirements are to be interpreted by having the deontic operator take wide scope over *some* conditional:[4]

Closure-W: $O(B(p) \rightarrow B(q))$.

Closure-W is not a conditional, and so it does not have the bootstrapping problem of Closure-N. But Closure certainly looks like a conditional, and so it would be preferable to have an interpretation according to which that is what it is. But even if we waive this constraint there are problems with Closure-W. What kind of conditional is embedded in Closure-W? It cannot be a material conditional. Suppose, to begin with, that "O" is closed under logical implication. In that case, the mere fact that you ought not to believe a proposition will entail that your believing that proposition normatively requires you to believe any other proposition, and the mere fact that you ought to believe a proposition will entail that your believing any other proposition normatively requires you to believe the first one. Although Broome does not consider this objection, he does consider a closely related one based on the assumption that logically equivalent propositions can be

[4] The reason for the emphasis on "some" will become apparent shortly.

substituted within the scope of "O." In that case, if you ought to believe a proposition *P*, then you have a normative requirement to believe *P* if a tautology is true. Thus, for instance, if you ought to believe that you have hands, then that either snow is white or snow is not white normatively requires you to believe that you have hands. Broome then concludes that the conditional embedded in Closure-W is not a material conditional, but a material conditional "with determination added, from left to right" (Broome (1999), p. 402).

Other authors, such as Kolodny (2005), have complained that wide-scope interpretations of normative requirements do not capture the "directionality" of those requirements. For instance, you can satisfy Closure-W just as well by ceasing to believe that *P* as you can by believing that *Q*. But, according to Kolodny, requirements such as Closure can only be satisfied by reasoning from the content of one attitude to another. If this is so, then Closure cannot be satisfied by ceasing to believe that *P* when one does not believe that *Q*. For when one does not believe that *Q*, there is no attitude (and, therefore, no content of an attitude) that one can reason from. Of course, one can reason from the fact that one does not believe that *Q* to the conclusion that one should give up one's belief that *P*. But one need not believe that one does not believe that *Q* when one does not, and so this content is not guaranteed to be available whenever one believes that *P* but fails to believe that *Q*. It might be argued that Kolodny's complaint misses the point that, according to Broome, the conditional embedded in Closure-W has "determination added, from left to right." In any case, whether within a wide-scope framework or outside of it, it seems clear that normative requirements should also satisfy this directionality requirement.

At a sufficiently general level, then, the puzzle of normative requirements is the challenge of finding an interpretation of normative requirements that satisfies all of the following constraints: (a) it avoids bootstrapping problems; (b) it respects their conditionality; and (c) it respects their directionality. Notice that, to satisfy (a), an interpretation must have it that detachment fails for normative requirements (i.e., that we cannot conclude that its consequent is true on the basis of the truth of its antecedent), whereas to satisfy (b) it must have it that normative requirements are conditionals. It follows that an interpretation satisfies both (a) and (b) only if it is an interpretation according to which detachment fails for conditionals—that is to say, an interpretation according to which Modus Ponens has counterexamples. I argue that there is an interpretation that satisfies all of the constraints, and I embrace the consequence that Modus Ponens has counterexamples. Moreover, my rejection of Modus Ponens is more radical—but better justified, I will argue—than recent similar rejections by, for instance, Dowell (2012) and Kolodny and MacFarlane (2010). My interpretation of normative requirements will also vindicate Fumerton's thesis and the principle of inferential justification, for satisfying a normative requirement, according to my interpretation, is not sufficient for the attitude mentioned in in its consequent to be justified.

7.3 Chisholm's Puzzle of Contrary-to-Duty Obligations

In 1963, *Analysis* published a revolutionary short paper that spawned a decades-long discussion and whose influence can still be felt today. I am referring, of course, to Roderick Chisholm's "Contrary-to-Duty Imperatives and Deontic Logic" (Chisholm (1963)). The literature generated by that paper is rich in logical sophistication and philosophical insight. I will argue that there is an intimate, but so far unnoticed, connection between the puzzle of normative requirements and Chisholm's puzzle of ctd obligations. Indeed, the key to finding an interpretation of normative requirements that satisfies (a) through (c) is to be found in an examination of Chisholm's puzzle.

Chisholm's puzzle concerns deontic logic. To the language of propositional logic we add the propositional operator O from section 7.2. The result is the language $\mathcal{D}$ of deontic logic. We can then give a Kripke-style possible world semantics for $\mathcal{D}$. A model for $\mathcal{D}$ will include a set of possible worlds W. In the usual presentation, models will also have a relation R over W, which is then informally interpreted as giving us, for each world $w \in W$, the set of worlds that are "accessible from" w (the set $\{w' \in W | wRw'\}$). We can then define a necessary truth at a world w as one that is true in every world accessible from w. But I will use an alternative presentation.[5]

According to this alternative presentation, in addition to the set W, models have a world-relative order $\leq_w$. It is generally assumed that $\leq_w$ is reflexive, transitive, and connected in W. The idea is that, for any world $w \in W$, $\leq_w$ ranks all the possible worlds according to how closely they match the ideal set by w—how good they are according to w (if $w' \leq_w w''$, then w' is closer to the ideal established by w than w'' is). Of course, w itself need not match that ideal particularly well, as it surely does not for the actual world (as we will see later, this is captured in the claim that deontic orders need not satisfy what Lewis called the "centering" assumption). The difference with the more usual presentation is that, instead of taking the set of accessible worlds as primitive, we construct them out of this order: the worlds accessible from a world w are those that are closest to the ideal set by w, according to $\leq_w$ (i.e., the set $\{w' \in W | \neg \exists w'' \in W w'' \leq_w w'\}$). We can then say that a proposition of the form $O(\phi)$ is true at a world w if and only if ϕ is true in all the best (i.e., accessible) possible worlds according to $\leq_w$. We are interested here in truth simpliciter, truth relative to the actual world. Thus, a proposition of the form $O(\phi)$ is true just in case ϕ is true in every possible world that matches the actual ideal (those are what I will call "perfect worlds"). This alternative presentation of the semantics for deontic logic will prove useful later in modeling the interaction of conditionals with modals.

[5] My alternative presentation follows Kratzer (1991b), who in turn is generalizing an idea from Lewis (1973).

In "Contrary-to-Duty Conditionals," Chisholm presented the following kind of scenario as a problem for $\mathcal{D}$ so interpreted. Suppose that:

1. You ought to take the trash out.

And suppose also that:

2. It ought to be that if you take the trash out, then you tell your spouse that you will do it.

But:

3. If you do not take the trash out, then you ought not to tell your spouse that you will do it.

Regrettably:

4. You do not take the trash out.

3 is what can be called a ctd conditional. Chisholm's puzzle starts from the idea that even if we neglect certain of our duties we still have obligations, for not all ways of violating one's duties are equally wrong. One feature that makes Chisholm's puzzle particularly interesting is that telling your spouse is obviously incompatible with not telling your spouse, and yet there seem to be good arguments that you have an obligation to do each (more on this below).

Chisholm's challenge is to find a formalization of 1 through 4 that preserves their mutual independence and consistency. 1 and 4 should be formalized in $\mathcal{D}$ as follows (where "t" stands for the proposition that you take the trash out):

1′. $O(t)$.

4′. $\neg t$.

Still within $\mathcal{D}$, 2 and 3 can be formalized as either material conditionals with a deontically modified consequent or as deontically modified material conditionals (where "s" stands for the proposition that you tell your spouse):

2′. $t \rightarrow O(s)$.

2″. $O(t \rightarrow s)$.

3′. $\neg t \rightarrow O(\neg s)$.

3″. $O(\neg t \rightarrow \neg s)$.

The grammar of 2 and 3 suggests that they should be interpreted as 2″ and 3′, but we need to consider all four possible formalizations of 1 through 4:

A	B	C	D
1′. $O(t)$	1′. $O(t)$	1′. $O(t)$	1′. $O(t)$
2′. $t \to O(s)$	2″. $O(t \to s)$	2″. $O(t \to s)$	2′. $t \to O(s)$
3′. $\neg t \to O(\neg s)$	3″. $O(\neg t \to \neg s)$	3′. $\neg t \to O(\neg s)$	3″. $O(\neg t \to \neg s)$
4′. $\neg t$	4′. $\neg t$	4′. $\neg t$	4′. $\neg t$

The problem is that all four formalizations run afoul of Chisholm's requirements. Formalization A violates the independence constraint, for 4' entails 2' (this reflects one of the problems that we mentioned for interpreting Closure as Closure-N). Formalization B also violates the independence constraint, for 1' entails 3": if in every perfect world t is true, then it follows that in every perfect world either $\neg t$ is false or *not–s* is true (this reflects one of the problems that we mentioned for interpreting Closure as Closure-W). Formalization C violates the consistency requirement. Notice that 3' and 4' entail $O(\neg s)$, but 1' and 2" entail $O(s)$ (if in every perfect world t is true and in every perfect world either t is false or s is true, then in every perfect world, s is true). But $O(\phi) \to \neg O(\neg \phi)$ is a theorem schema of deontic logic[6] (and is, in any case, something that we think is true). Finally, formalization D inherits the problems of both A and B. Therefore, there is no formalization of 1 through 4 in $\mathcal{D}$ that respects both the consistency and the independence constraint. Moreover, notice that all formalizations except A make at least one of 2 and 3 turn out not to be a conditional.

It is interesting to note that different authors have insisted on imposing additional constraints beyond the independence and consistency of 1 through 4. Some authors (DeCew (1981), Lewis (1974)) require that sentences like 1 and 2 entail a sentence like:

5. $O(s)$.

Some other authors (Mott (1973), Chellas (1974), al-Hibri (1978)) require that sentences like 3 and 4 entail a sentence like:

6. $O(\neg s)$.

6 is what can be called a ctd obligation. Following Greenspan (1975), let us call the first constraint the requirement of "deontic detachment" and the second one the requirement of "factual detachment." Notice that an interpretation better not satisfy both deontic and factual detachment, for that would render 1 through 4

[6] Although only if the order has a maximum, for without maximum there are no guarantees that there are accessible worlds, and without accessible worlds the schema fails.

inconsistent given that incompatible obligations are impossible. The argument for deontic detachment is clear and convincing: you ought to take the trash out and so tell your spouse, and the mere fact that you will not take the trash out does not make these obligations disappear. The argument for factual detachment turns essentially on the claim that not all ways of sinning are equally bad. Fans of factual detachment assume that we can capture this obvious truth only by claiming that sinners do not have the same obligations as non-sinners. I argue in section 7.5 that, on the contrary, we can distinguish better and worse ways of sinning without claiming that sinning changes what obligations you have. If that argument works, then we can hold on to deontic detachment without doing violence to the grain of truth behind factual detachment.

But although deontic detachment is incompatible with factual detachment, some proponents of deontic detachment have sought to capture a different kind of factual detachment—"strong factual detachment" (Loewer and Belzer (1983), Prakket and Sergot (1997), Kolodny and MacFarlane (2010)). Strong factual detachment allows us to conclude 6 not from 3 and 4, but rather from 3 and a strengthening of 4 according to which it is not only the case that you will violate your obligations, but it is somehow "settled" that you will do so.[7] One kind of argument for strong factual detachment is that it is supposed to enshrine some kind of "ought implies can" principle, for if it is settled that you will not fulfill your unconditional obligation, then allegedly you cannot fulfill it and so it is not the case that you ought to fulfill it, which allows the ctd obligation to "kick in." I discuss whether we should adopt strong factual detachment in addition to deontic detachment in section 7.7.

Some authors (Greenspan (1975), Loewer and Belzer (1983)) have thought that paying close attention to the times at which the obligations are had is crucial for solving the problem. But this solution can't work in general, because the time of the ctd obligation can be the same as the time of the unconditional obligation.[8]

Finally, there is the question of whether 2 and 3 should receive the same interpretation (as in A and B) or different ones (as in C and D). The only substantive difference between 2 and 3 is whether there is an obligation to bring about their antecedents, which is a difference that is not intrinsic to 2 and 3 themselves. Therefore, I adopt the requirement that 2 and 3 be given the same interpretation.

A solution to Chisholm's puzzle of ctd conditional obligations, then, must satisfy the following requirements: (1) it must render propositions such as 1 through 4 mutually compatible and independent, and (2) it must make them

[7] For MacFarlane and Kolodny, for instance, 4 has to be known—or, more precisely, it has to be part of the informational set relative to which deontic sentences are evaluated—in order for 6 to follow.

[8] As in the gentle murder paradox of Forrester (1984) and, more importantly for my purposes, the case of normative requirements.

satisfy deontic detachment. The language of $\mathcal{D}$ just doesn't have enough resources to satisfy the first of these requirements.

7.4 A Solution to Chisholm's Puzzle

It is by now commonly accepted that Chisholm was right that standard deontic logic cannot capture the kind of conditional obligation at issue in his puzzle. There is no agreement, however, on how to modify standard deontic logic in order to capture those kinds of conditional obligations. Some authors suggest that such obligations must be understood in terms of a conditional different from the material conditional, whereas others advocate the introduction of a primitive dyadic operator. In this section I briefly consider both strategies and side with those who favor the introduction of a dyadic operator.

As an instance of the first approach to the problem, some authors (Mott (1973), Chellas (1974), al-Hibri (1978)) propose to augment the language of $\mathcal{D}$ with a counterfactual conditional (for which we will use '$\Rightarrow$'), and claim that 2 and 3 are to be interpreted in terms of counterfactuals. Let us call the resulting language $\mathcal{DC}$. Semantics for counterfactuals also use Kripke models. Formally, a semantics for counterfactuals along the lines of Stalnaker and Lewis is very similar to the semantics for $\mathcal{D}$ that we discussed in section 7.3. Still, there are two important differences. First, the order relevant to the interpretation of counterfactuals is not supposed to be deontically based, but rather based on a relation of overall similarity. Second, whereas a deontically modified proposition is semantically analyzed as a universal quantification over all the perfect worlds, a counterfactual is analyzed as a universal quantification over the most similar worlds where the scope of the quantifier is restricted by the antecedent. A counterfactual proposition, $(\phi \Rightarrow \psi)$, is true if and only if there is some world x where the antecedent and the consequent are both true, and x is closer (according to the overall similarity ordering) to the actual world than any world where the antecedent is true and the consequent is false. So, whereas models for $\mathcal{D}$ have a single deontic order over the set of possible worlds, models for $\mathcal{DC}$ have two orders: a deontic one and one based on a relation of overall similarity.

Could we use $\mathcal{DC}$ to solve Chisholm's puzzle? The language now allows us to interpret conditional obligations in these two new ways:

7. $\phi \Rightarrow O(\psi)$

8. $O(\phi \Rightarrow \psi)$

Take 7 first. Interpreting conditional obligations as counterfactuals with a deontically modified consequent does not have all the same problems as interpreting them as material conditionals with a deontically modified consequent. For instance, as far as satisfying Chisholm's requirements of consistency and independence goes, taking conditional obligations to have the form of 7 does solve the puzzle. However, it is easy to see that if we do this then we will have satisfied factual detachment and violated deontic detachment. Counterfactuals obey Modus Ponens (I explain why shortly), and so the interpretation has the consequence that if the antecedent of a conditional obligation is true then the consequent must be true. Thus, interpreting conditional obligations as in 7 will not do.

How about interpreting them as in 8? That interpretation also satisfies the consistency and independence requirement. Moreover, the interpretation satisfies deontic detachment. As far as the formal constraints go, then, that interpretation is perfect. Unfortunately, it won't do either.

To begin with, notice that counterfactuals are usually taken to entail the corresponding material conditional. Therefore, some of the objections that we leveled against Closure-W apply to interpreting conditional obligations as in 7. For instance, for any ϕ such that we have an unconditional obligation to ϕ, the counterfactual that if T (an arbitrary tautology) were true then we ought to ϕ is true. That is to say, if we interpret conditionals obligations as in 7, then for any unconditional obligation ϕ we have a conditional obligation to ϕ if a tautology is true.

Maybe some will not find that consequence so unpalatable. But it gets worse. A deontic proposition is true just in case the proposition that it modifies is true in all the perfect worlds, and a counterfactual proposition is true just in case there is a world where both the antecedent and the consequent are true that is closer to the actual world than any world where the antecedent is true but the consequent is false. So, taken together, these clauses tell us how to determine whether a deontically modified counterfactual is true. We can think of it as a two-stage process. First, isolate the set of perfect worlds, a set determined by the deontic ordering. Second, for every one of those worlds *w*, figure out whether some world where both the antecedent and the consequent of the counterfactual are true is closer (according to the overall similarity ordering) to *w* than any world where the antecedent is true but the consequent is false. The problem for the interpretation of conditional obligations as deontically modified counterfactuals is that the deontic and the overall similarity ordering need not intersect each other in the way that is required to satisfy our intuitions about which propositions are true. That is to say, it may well be that a world where things are very bad is overall more similar to a perfect world than every world where things are somewhat bad, in which case an intuitively true conditional obligation will turn out to be false.[9]

[9] DeCew (1981) seems to have something like this objection in mind.

Let us take as our example of a conditional obligation the proposition that if you sin you ought to repent. Now, according to the interpretation under consideration, this proposition should be read as: it ought to be the case that if you were to sin you would repent. In every perfect world, you do not sin, of course. Here are two crucial questions. First, is it true that in every perfect world you have the disposition to repent if you sin? Second, is it true that every world where you have the same dispositions as in *w* is closer to *w* than any world where you don't? I will argue that if the answer to either of these questions is "No," then we cannot interpret conditional obligations as deontically modified counterfactuals, and that there are good reasons to think that the answer to both questions is "No."

Take the first question first. Some people may think that in every perfect world not only do you not sin, but, moreover, you have a stable disposition to repent if you sin. I have my doubts about this. It may well be that if you have a stable disposition not to sin but no disposition to repent (perhaps you don't even have the capacity to repent) then everything is as it should be as far as you are concerned, even if you would not repent were you (very improbably) to sin. Suppose, then, that there are at least some perfect worlds where you do not have the disposition to repent if you sin. In at least some of those worlds, then, the counterfactual that if you were to sin you would repent is false. For, in at least some of those worlds, a world where you sin but retain your dispositions (and so you do not repent) will be closer to that perfect world than any world where you sin and do not retain your dispositions.

But suppose that it is true that in every perfect world you have the disposition to repent if you sin. Still, for the conditional obligation to repent if you sin to be true according to the interpretation under consideration, it has to be the case that for every perfect world *w*, a world where you sin and retain your dispositions (and so repent) is closer to *w* than a world where you sin but do not repent (and so do not retain your dispositions). But surely in at least some perfect world this is not true. For consider the fact that in at least some perfect worlds you are a saint: not only do you not sin and have the disposition to repent if you sin, but it would be extremely hard for you to sin. In particular, it may be the case that you would sin only if you were a very different person, with very different dispositions. Therefore, if (unthinkably) you were to sin, you would not repent. But then, obviously, the counterfactual that if you were to sin then you would repent is false.

I conclude, then, that even though introducing counterfactuals allows us to solve Chisholm's puzzle in the sense that it allows us to give interpretations of 1 through 4 that make them come out as consistent and independent of each other, those interpretations are not satisfactory for other reasons. First, interpreting 2 and 3 as counterfactuals with deontically modified consequents means that 2 and 3 satisfy factual detachment, and so that they do not satisfy deontic detachment. Interpreting them as deontically modified counterfactuals fares better in this regard, for under that interpretation 2 and 3 satisfy deontic detachment and violate factual detachment. But the truth-conditions of deontically modified counterfactuals need not coincide with the truth-conditions of propositions like 2 and 3.

I turn now to consideration of the approach that favors introducing a primitive dyadic operator.[10] According to this approach, instead of adding a monadic deontic operator to the language of propositional logic, we add a dyadic deontic operator: $O(-|-)$.[11] We can then define a monadic operator as follows:

$$O(\phi) =_{def.} O(\phi|T)$$

where "T" is any tautology. Instead of having two orders over W, like the models for $\mathcal{DC}$, models for dyadic deontic logic have only the deontic one. A conditional obligation proposition $O(\phi|\psi)$ is true just in case ϕ is true in all the best worlds where ψ is true.[12] Notice that, formally, dyadic conditional obligations have exactly the same semantics as counterfactuals.[13] Thus, whereas unconditional deontic propositions are analyzed as universal quantifiers over perfect worlds (in both monadic and dyadic deontic logic), conditional deontic propositions are analyzed in dyadic deontic logic as universal quantifiers over the best possible worlds where the scope of the quantifier is restricted by the "antecedent," just like counterfactuals.

Still, there are of course important differences between counterfactuals and dyadic conditional obligations. The order for counterfactuals is based on a relation of overall similarity instead of being deontic, for instance. Moreover, the overall similarity ordering relevant for the analysis of counterfactuals satisfies the centering assumption,

Centering: For every world w, no world is more similar to w than w itself is,

whereas, as I remarked above, the deontic ordering does not, in general, satisfy centering—many worlds are not perfect, of course, and many of them are not perfect even according to themselves. Interestingly, the fact that the ordering relevant to the analysis of counterfactuals satisfies centering is responsible for the fact that counterfactuals entail the corresponding material conditional, which in turn is responsible for the fact (already alluded to) that they validate Modus

[10] This approach was pioneered by Wright (1956), who developed it in response to Prior (1954). Authors who think that this approach provides the best response to Chisholm's puzzle include Danielson (1968), Hansson (1969), and Lewis (1973).

[11] Notice that, in this formulation, the "antecedent" is on the right.

[12] More carefully, to avoid the limit assumption: the conditional obligation proposition is true if and only if there is some world w where both ψ and ϕ are true, and w is closer to the actual world than any world where ψ is true and ϕ is false.

[13] For details, see for instance Lewis (1974).

Ponens. From a syntactic point of view, dyadic conditional obligations are no more conditionals than statements of conditional probability are conditionals.[14] Still, we may wonder whether dyadic conditional obligations validate Modus Ponens in the following sense: is $O(\phi)$ guaranteed to be true at a world w whenever both ψ and $O(\phi|\psi)$ are true at w? The answer is that, given that the deontic ordering does not satisfy centering, dyadic conditional obligation does not validate Modus Ponens in this sense. Thus, for instance, whereas all the best worlds where you sin are worlds where you repent, all the perfect worlds are worlds where you do not sin (and so do not repent), even if you actually do sin. In general, the best ϕ-worlds need not be the same as the best worlds according to every ϕ-world. For instance, the best worlds where you sin are not at all like the best worlds according to worlds where you sin. The best worlds where you sin are worlds where you repent, but the best worlds are worlds where you do not sin and so do not repent, and this is so even according to worlds where you do sin. Dyadic conditional obligations do satisfy deontic detachment, though. For suppose (for reductio) that $O(\psi)$ and $O(\phi|\psi)$, but not $O(\phi)$. Then, by the first and third assumptions, there is some perfect world where ψ is true but ϕ is false. But then it is not true that every perfect world where ψ is true is one where ϕ is also true, which contradicts the second assumption.

Interpreting conditional obligations as dyadic conditional obligations, then, satisfies Chisholm's requirements of independence and consistency. In addition, conditional obligations so interpreted satisfy deontic detachment and do not satisfy factual detachment. Moreover, because only a deontic ordering is at issue in the models for dyadic deontic logic, there is no danger that that order will not accord with our judgments about which propositions are true. If need be, we can just change the order to suit our judgments (more on this in section 7.5). I conclude, then, that dyadic deontic logic, with its primitive notion of conditional obligation, gives us a satisfactory solution to the problem of ctd conditional obligations.

7.5 Normative Requirements Revisited

Now let us return to the problem of how to interpret Closure and other normative requirements. Recall that interpreting it as Closure-N led to the problem of bootstrapping (among others), and interpreting it as Closure-W led to the problem of lack of directionality (among others). We can now apply to that problem the lessons learned in examining Chisholm's puzzle.

[14] Although it could be argued that it is a different kind of conditional, just like subjunctive conditionals are different from indicative ones. The fact that counterfactuals are also formalized as dyadic operators is relevant here.

First, it will not do to interpret Closure in either of these ways:

Closure-CN: If it were the case that you believe that p, then it would be the case that you ought to believe q.

Closure-CW: It ought to be that, if it were the case that you believe that p, then it would be the case that you believe q.

The problem with Closure-CN is that counterfactuals validate Modus Ponens, and so Closure-CN leads to the bootstrapping problem as much as Closure-N does. Thus, just because you irrationally believe that it is raining when you have absolutely no evidence for that proposition, Closure-CN has it that you ought to believe that it is raining. But could it be that the problem comes from focusing on counterfactuals, and some other kind of conditional will work? Indeed, I will argue that this is so. But bear in mind that, as long as the conditional in question satisfies Modus Ponens, the interpretation will have the bootstrapping problem.

What about Closure-CW? Notice, first, that Closure-CW does not satisfy constraint (b): according to this interpretation, 2 and 3 turn out not to be conditionals. Moreover, notice that, using our same example, Closure-CW entails the following:

9. It ought to be that, if you were to believe that it is raining, then you would believe that it is precipitating.

Is 9 true? It's not clear. We have to first consider all those worlds that are epistemically perfect (and that are not otherwise gratuitously different from this world). In none of those worlds do you believe that it is raining. Is it true that, for every one of those worlds w, if there is a world w' where you believe that it is raining but do not believe that it is precipitating, then there is another world w'' where you believe both propositions and w'' is closer to w than w' is? I don't see why we should believe that. It may be that worlds where you believe that it is raining are so different from the epistemically best worlds that in those you do not believe it is precipitating. And it may be that, although you do not believe that it is raining and have a strong disposition not to believe it, if you were to believe it then you would be such a different person that you would not believe that it is precipitating.

But what about alternative wide-scope interpretations? Maybe taking normative requirements to be obligations with wide scope over a counterfactual does not work, but changing the kind of conditional in question does? This is, remember, Broome's strategy. He holds that normative requirements are obligations with wide scope over a conditional that is like a material conditional but "with determination added from left to right." But I do not think that the kind of conditional matters in this case. I will argue that the kind of conditional does matter for Closure- N, but the problem for Closure-W stems fundamentally from the fact that the semantics for the conditional, of whatever kind, is not guaranteed to march in lockstep with the semantics for deontic modals. Changing the kind of conditional may change the kind of conflict that arises, but we have no good reason to think that the conflict will be avoided. By contrast, as we shall see, we do have a very important reason to think that the conflict is avoided when the obligation takes narrow scope over the conditional.

But we have not run out of possible interpretations. What about a dyadic interpretation of normative requirements? According to this interpretation, we rank all the possible worlds according to how epistemically good they are, and a sentence like:

10. O(B(it is precipitating)|B(it is raining))

is true if and only if all the epistemically best worlds where you believe that it is raining are worlds where you believe that it is precipitating ("B" is here a belief operator).

Broome himself called attention to the notion of conditional obligation, but only to dismiss it as an interpretation of normative requirements. His reason for the dismissal, however, is puzzling. Here is what Broome says:

> Deontic logic contains a notion of conditional obligation. . .which could serve as a model for normative requirements. . .But deontic logic will not give us much help because the analysis of conditional obligation remains unsettled.
>
> (Broome (1999), p. 403, n. 3)

I am not sure what Broome means when he says that "the analysis of conditional obligation remains unsettled." He might be referring to the problem of how to account for ctd conditional obligations. In that case, although it is true that there are interpretations of ctd conditional obligations that are competitors to the dyadic interpretation, I argued against them in section 7.4. But he might also be referring to the interpretation of the notion of conditional obligation in dyadic deontic logic itself. Although such a notion has a standard semantics, Broome may be hinting that the standard semantics is not up to the philosophical task of yielding an interpretation of normative requirements (or conditional obligations in general). In any case, whether this is Broome's worry or not, I now turn to it.

The worry can be put starkly in these terms: what does epistemic goodness, on the basis of which the worlds are ranked, consist in? It is not clear that, presented with a description of two worlds, we can make sense of the question of which one of them is epistemically better. Are we being asked in which world more propositions are known? Or in which world more propositions are rationally believed? Or which world contains more epistemic virtue? Insofar as the semantics itself doesn't answer this question, it is useless as a philosophical interpretation of conditional obligation. But although this worry is natural, it is misguided.

Consider a similar worry that has been raised against the Lewis-Stalnaker possible-worlds semantics for counterfactuals. As we have seen, according to that semantics a counterfactual is true at a world *w* just in case its consequent is true in the closest worlds to *w* where the antecedent is also true. Thus, the semantics for counterfactuals includes an ordering of possible worlds in terms of a relation of overall similarity. Now Kit Fine has worried that this relation of overall similarity will not track our judgments about which counterfactuals are true (Fine (1975)). Thus, to take Fine's example, a world where Nixon presses the button but the bomb doesn't go off is surely more similar to the actual world than a world where the bomb does go off and a nuclear war ensues. But, still, we think that if Nixon had pressed the button the bomb would have gone off. Therefore, the objection goes, the possible-worlds semantics for counterfactuals fails.

Notice, first, that my objection to understanding normative requirements in terms of counterfactuals (or any other conditional) does not mirror Fine's objection to the semantics of counterfactuals themselves. My objection is simply that the truth-values of those conditionals are not guaranteed to correspond to the truth-values of normative requirements (for instance, I plausibly have a normative requirement to believe that it is precipitating given that I believe that it is raining, but it is arguably false that it ought to be the case that if I were to believe that it is raining then I would believe it is precipitating).

It is a good thing that my objection does not mirror Fine's, because I think that it is misguided. In answering this objection, Lewis wrote:

> The thing to do is not to start by deciding, once and for all, what we think about similarity of worlds, so that we can afterwards use these decisions to test [the analysis]. Rather, we must use what we know about the truth and falsity of counterfactuals to see if we can find some sort of similarity relation—not necessarily the first one that springs to mind—that combines with [the analysis] to yield the proper truth conditions. (Lewis (1979), p. 43)

Lewis's idea here is that a semantics is not a heuristic. Although Lewis's project is a reductive one, in the sense that he wants to analyze modality in terms of possible worlds and a similarity relation between them, the order of discovery need not (and does not) match the metaphysical order. We do not discern which worlds are more similar to which ones by using a telescope and inspecting the order of the worlds in metaphysical space. Rather, we use our knowledge of which counterfactuals are true and which ones false to answer questions about which worlds are closer to which.

Analogously, the semantics for conditional obligations is not a heuristic either. The ranking of worlds according to how good they are is not an independent yardstick that issues verdicts about the truth-values of normative requirements that we can compare with our own unaided verdicts. A world where you believe both that it is raining and that it is precipitating is epistemically better than a world where you believe the former but not the latter if and only if there is a normative requirement to believe that it is precipitating given that you believe that it is raining. But, importantly, questions about the left-hand side of this biconditional are to be answered in terms of its right-hand side, and not vice versa.

This does not mean, however, that the semantics is idle, any more than it means that the possible-world semantics for counterfactuals is idle. On the contrary, as we shall soon see, interpreting normative requirements as conditional obligations has important and interesting consequences for the logic of normative requirements, and opens up the possibility of resolving the issues that beset other interpretations.

How does this interpretation, then, fare with respect to our constraints? It satisfies constraint (a): it avoids bootstrapping problems. It does so in virtue of failing to satisfy factual detachment. But isn't this a problem? If normative requirements do not satisfy factual detachment, in what sense are they normative? They do not entail that we have any unconditional obligation, not even when their antecedent is satisfied. If a requirement says that we should believe a certain proposition under certain conditions and those conditions are satisfied, shouldn't it follow that we ought to believe the proposition in question?

My reply is that we should take to heart the lessons learned in considering Chisholm's puzzle. Normative requirements are conditional obligations, and conditional obligations satisfy deontic detachment. Moreover, no consistent operator can satisfy both deontic and factual detachment. Therefore, normative requirements cannot satisfy factual detachment—which means that they cannot entail unconditional obligations even when their antecedent is satisfied. Does this mean that normative requirements are not, after all, normative? If a requirement is normative only if it entails unconditional obligations when its antecedent is satisfied, then the answer is, obviously, that normative requirements are not normative. This result, however, shouldn't be seen as an objection to the account of normative requirements I propose. It is, rather, an indication that normative requirements are ill-named—which is, in turn, an indication that the proper nature of those requirements is ill-understood in the literature.

What about the other two requirements? Does the interpretation of normative requirements in terms of dyadic conditional obligations respect the conditionality of normative requirements (constraint (b))? Does it respect their directionality (constraint (c))? Let us start with directionality.

The obvious unpacking of the metaphor of directionality is that the normative requirement to ψ given that you ϕ may only be satisfied in one way: by ψ-ing. Now in "Why Be Rational," Kolodny seems to think that to respect the directionality intuition one must give an interpretation of normative requirements that satisfies factual detachment. But if this is what he thinks, he is wrong about that, for the directionality intuition is not about the truth-conditions of normative requirements, but rather about their satisfaction conditions (and the same goes, of course, for conditional obligations in general). Thus, the interesting question is not whether the interpretation of normative requirements as dyadic conditional obligations itself satisfies the directionality intuition, but rather whether it is compatible with an account of their satisfaction conditions that respects that intuition.

Let us say that if you have a normative requirement to ψ given that you ϕ, then you satisfy that requirement only if you ψ and you violate it if you ϕ but don't ψ. Now, satisfying a normative requirement does not entail that you will discharge any obligations you have—indeed, given that some normative requirements have ctd obligations in their "consequent," satisfying a normative requirement may entail that you fail to discharge obligations you have. So, for instance, since you have a normative requirement to believe that it is precipitating given that you believe that it is raining, you satisfy that requirement only if you believe that it is precipitating. But suppose that you ought not believe either that it is raining or that it is precipitating. We seem, then, to have a conflict. If you don't believe that it is precipitating, then you do not satisfy the normative requirement, but if you do then you don't satisfy your unconditional obligations. But there is no real conflict, because there are two ways of not satisfying a dyadic conditional obligation (and so a normative requirement, under the interpretation that we are considering): you may violate it, but you may also cancel it.

Compare dyadic conditional obligations with conditional bets.[15] Suppose that you think that if you win the lottery you will be happy. I think that you are wrong: winning the lottery will actually make you unhappy. It will not be wise of me to bet against the material conditional, which is equivalent to the following proposition: either you do not win the lottery or you are happy. That proposition is extremely likely to be true, just because it is extremely likely that you will not win the lottery. But it may be wise of me to propose a conditional bet: a bet on whether you are happy, conditional on your winning the lottery. If you win the lottery and you are happy, I lose the bet. If you win the lottery and you are unhappy, I win the bet. In the most likely case where you do not win the lottery, the bet is off: neither one of us owes the other anything.

Similarly, nothing stops us from conceiving of the satisfaction conditions of dyadic conditional obligations in the same way. If you ϕ and ψ, you satisfy your conditional obligation. If you ϕ but don't ψ, you violate your conditional obligation. But if you don't ϕ, then you neither satisfy nor violate your conditional

[15] See Michael Dummett (1959), p. 8 on conditional bets, and Willard van Quine (1950) on conditional assertions. See also Kris McDaniel and Ben Bradley (2008).

obligation. Your conditional obligation is simply cancelled. You can avoid violating conditional obligations of this kind in two ways: by satisfying them or by canceling them. And when the conditional obligation is ctd, you can only discharge your unconditional obligations by canceling the conditional one.

This account of the satisfaction conditions of conditional obligations is compatible with their interpretation as dyadic conditional obligations. Therefore, satisfying the directionality condition does not mean that normative requirements satisfy factual detachment. On the contrary, they satisfy deontic detachment. So, if you believe that it is raining without believing that it is precipitating, then you violate a normative requirement. That does not mean that you ought to believe that it is precipitating. You ought not to believe that it is raining, and, given your other obligations, you ought not to believe that it is precipitating. So, what ought you to do (from an epistemic point of view) in that situation? You ought not to believe either proposition, thus satisfying all your unconditional obligations and avoiding violating any conditional obligation. It is still true, however, that the only way to satisfy your conditional obligation is by believing that it is precipitating as well as that it is raining.

Even if the interpretation of normative requirements as dyadic conditional obligations has all the advantages that I have claimed for it, there is still the question whether it satisfies constraint (b). Are dyadic conditional obligations conditionals? There are two related reasons to worry that they are not. First, they do not satisfy factual detachment, which is to say that Modus Ponens is not valid for them. But what are conditionals good for if not for detachment? Second, if the dyadic conditional obligation interpretation is correct, then the "if" clauses in 2 and 3 are idioms, because the meaning of 2 and 3 cannot be captured by the separate components "if" and "ought." In "Ifs and Oughts," MacFarlane and Kolodny argue against the dyadic approach to conditional obligations precisely along the lines of this second objection. They then go on to present an alternative picture—an assessment-relative account of deontic (and epistemic) modals married to a view of conditionals as restrictors. In the section 7.6 I defend the dyadic approach to conditional obligations from MacFarlane and Kolodny's objections, and in section 7.7 I criticize their own alternative account.[16]

7.6 The Restrictor View of Conditionals

MacFarlane and Kolodny's main objection to the dyadic approach is that it makes "if. . .ought" constructions come out as idioms: their meaning is not compositionally determined by the meaning of "if" together with the meaning of "ought."

[16] MacFarlane and Kolodny's target is neither Chisholm's puzzle nor the problem of interpreting normative requirements, but rather the miner's puzzle that Parfit takes from Reagan (1980).

It is true that the truth-conditions for dyadic conditional obligations are not arrived at compositionally, from a semantics for "if" and one for "ought." But notice, to begin with, that the same happens for the standard Stalnaker-Lewis account of counterfactuals as variably strict conditionals. In any case, it is possible to obtain the truth-conditions for conditional obligations compositionally, by adopting a view of conditionals that is widely held in linguistics (and which MacFarlane and Kolodny themselves adopt).

Lewis (1975) considered sentences like the following:

11. Usually, if Mary is happy, John is glad.

For our purposes, we can think of "usually" as a quantifier over times or situations, and so we can treat both "Mary is happy" and "John is glad" as being true or false relative to these times or situations. But what contribution does "if" make to the meaning of 11? We might think that it is some kind of connective joining together the two sentences "Mary is happy" and "John is glad" to produce another complex sentence which is itself true or false relative to times or situations. But Lewis argued that this is not the case. To take just one simple example, suppose that we take the conditional in 11 to be the material conditional. Suppose now that Mary is usually unhappy, but in those rare occasions where she is happy, John is not glad. In that case, 11 would turn out to be true—but this is the wrong result. Rather, Lewis argued, the entire "if" clause acts as a restrictor on the quantifier. That is to say, according to Lewis, we can paraphrase 11 as follows:

12. Most situations in which Mary is happy are situations where John is glad.

Kratzer (1991a) argued that Lewis's analysis of 11 can be extended to cases where conditionals interact with modals, as in the following examples:

13. Necessarily, if Mary is happy, John is glad.
14. If Mary is happy, John must be glad.
15. If Mary is happy, John might be glad.

According to Kratzer, those sentences can be interpreted (roughly) as follows:

16. All possible situations where Mary is happy are situations where John is glad.
17. All situations compatible with what we know where Mary is happy are situations where John is glad.
18. Some situations compatible with what we know where Mary is happy are situations where John is glad.

Moreover, Kratzer argued, even in the case of "bare conditionals," where the "if" clause doesn't explicitly interact with any modal, the sentences do contain an implicit modal operator. Thus, the unadorned

19. If Mary is happy, John is glad

is to be interpreted as 14, which in turns is to be interpreted as 17. Thus, Kratzer concludes,

> The history of the conditional is the story of a syntactic mistake. There is no two-place "if then" connective in the logical forms for natural language. "If"-clauses are devices for restricting the domains of various operators.
>
> (Kratzer (1991a), p. 11)

It is fair to say that the restrictor view is the received view in linguistics.[17] And notice that, according to this view, 2 and 3 should be interpreted as follows:

20. All the best worlds where you take the trash out are worlds where you tell your spouse.
21. All the best worlds where you do not take the trash out are worlds where you do not tell your spouse.

But 20 and 21 are just the truth-conditions one gets from the dyadic analysis.

7.7 Modus Ponens and Strong Factual Detachment

I said that ctd conditional obligations must satisfy deontic detachment—and so not satisfy factual detachment. Similarly, I said that normative requirements must be conditionals that do not detach (on pain of leading to bootstrapping problems), and so that they must violate Modus Ponens. I will now make good on my promise to argue that those are the right constraints.

The motivation for requiring factual detachment arises from the idea that not all ways of sinning are the same. It is one thing, for instance, to not take the trash out when it is your turn, but if in addition you told your spouse that you did, you are only adding insult to injury. I said before that we need not accept factual detachment in order to capture this motivation, and I am now in a position to explain why. We can capture the motivation by noticing that saying that you have a conditional obligation not to tell your spouse if you do not take the trash out entails that not taking the trash out and telling your spouse that you did is worse

[17] Von Fintel (2012) calls it "the dominant approach to semantics in linguistics."

than not taking the trash out and not telling your spouse. That is how we can have guidance for sinners without having factual detachment. If you don't take the trash out, you are still obligated to do it and tell your spouse. But if you don't take the trash out and tell your spouse, then you are actualizing a worse world than if you don't take the trash out and don't tell your spouse.

Similarly, I argued in section 7.5, the fact that you irrationally believe that it is raining does not entail that you ought to believe that it is precipitating, not even if it is true that if you believe that it is raining then you ought to believe that it is precipitating. At the level of pre-theoretic judgments, that you in fact believe that it is raining does not change the fact that you ought not believe it. Now suppose that, in fact, rain is the only kind of precipitation in your area (and you know this). Then, the fact that you ought not to believe that it is raining does entail that you ought not believe that it is precipitating. Given that you cannot have contradictory obligations, that you ought not to believe that it is raining entails that you ought not to believe that it is precipitating. At the level of the formal semantics, these judgments are captured in the fact that the deontic ordering is not centered—we do not require every world to consider itself among the perfect ones. If we now interpret normative requirements (and, more generally, conditional obligations) as dyadic conditional obligations—or, equivalently as far as the truth-conditions go, as conditionals whose antecedents have the job of restricting universal quantification over perfect worlds—then the failure of factual detachment follows. For, in interpreting normative requirements as dyadic conditional obligations, we interpret, for instance, the claim that believing that it is raining normatively requires you to believe that it is precipitating as saying that the best worlds where you believe that it is raining are worlds where you believe that it is precipitating. But this does not mean that the best worlds according to worlds where you believe that it is raining are worlds where you believe that it is precipitating. On the contrary, the best worlds according to worlds where you irrationally believe that it is raining (and where you know that rain is the only kind of precipitation in your area) are worlds where you do not believe that it is precipitating.

Some authors think that merely pointing out that a world where you believe both that it is raining and that it is precipitating is better than a world where you believe that it is raining without believing that it is precipitating is not enough. One motivation for saying this is the already examined claim that this robs normative requirements of their normativity. After giving a semantics in line with the one presented here, for instance, John Cantwell goes on to say the following (Cantwell is considering a case where Smith steals from John):

> But do we get the right answer? I do not know. On the one hand I think that as Smith steals from John, he ought to be punished; on the other hand, it ought to be the case that this is a world where Smith is not punished (as this ought to be a world where he does not steal). (Cantwell (2008), p. 348)

Cantwell himself doesn't provide a way for us to have our cake and eat it too, but other authors have done so. I consider here the proposals of MacFarlane and Kolodny and Janice Dowell.

In MacFarlane and Kolodny's semantics for deontic and epistemic modals, sentences are evaluated at possible world states and information states pairs, $<w, i>$, where a possible world state is an assignment of extensions to all the basic (i.e., non-deontic or epistemic) predicates and terms of the language, and an information state is a set of possible world states. For sentences without deontic or epistemic predicates, i is irrelevant to their truth. But $O(\phi)$ is true at $<w, i>$ iff for all $w' \in d(i)$, ϕ is true in $<w', i>$, and $E(\phi)$ (it (epistemically) must be the case that ϕ) is true at $<w, i>$ iff for all $w' \in e(i)$, ϕ is true in $<w', i>$—where d is the deontic selection function and e is the epistemic selection function. MacFarlane and Kolodny take the epistemic selection function to be the identity function. For the deontic selection function, they impose the constraint that it be realistic: $d(i)$ must be a subset of i. Informally, $e(i)$ returns the worlds that (epistemically) must be true at $<w, i>$, and $d(i)$ returns the ideal worlds at $<w, i>$. MacFarlane and Kolodny also hold that d can be seriously information-dependent: a world that is ideal relative to i need not be ideal relative to every subset of i.[18] As for the conditional, MacFarlane and Kolodny adopt Kratzer's restrictor picture (already presented).

Dowell follows Kratzer not only on conditionals, but also on the semantics of modal statements. On this view, modal statements are evaluated with respect to a modal base (the set of possible worlds on which the statement is to be evaluated) and a ranking of those worlds. Both the modal base and the ranking can be explicit in linguistic material or contextually provided by the speaker's intentions.

For Dowell as well as for MacFarlane and Kolodny, conditionals do not obey Modus Ponens either. In this, of course, I agree. But whereas their semantics invalidates Modus Ponens, it validates it in the special case where the premises are epistemically necessary—that is to say, their semantics satisfies what we earlier called strong factual detachment. However, the very same cases that motivate rejection of factual detachment also motivate rejection of strong factual detachment. It is not only true that if you do not take out the trash, then you ought not to tell your spouse that you did it—it is, we may assume, epistemically necessary that this is so (we know that it is true). And it is not only true that you will not take out the trash—it is also epistemically necessary. That these premises are epistemically necessary (for us) means that the information state i with respect to which we evaluate them includes only world states where they are true—and this will be so whenever we evaluate those sentences with respect to information states that capture what we know. But then, it follows that you ought not to tell your spouse

[18] MacFarlane and Kolodny introduce complications to their semantics in order to deal with modals in the antecedent of a conditional, but I ignore this complication because it is orthogonal to the issues I am interested in.

that you will take the trash out. That is to say, both Dowell's and MacFarlane and Kolodny's picture satisfy factual detachment for the special case where the premises are epistemically necessary.[19]

This result is unacceptable. The four sentences in Chisholm's puzzle are jointly satisfiable in their semantics,[20] but not with respect to an information set that presupposes the truth of the sentences in the set.[21] Thus, their semantics cannot solve Chisholm's puzzle if we assume that we are evaluating the sentences involved in the puzzle with respect to an information set that presupposes the truth of those sentences. I conclude, then, that Dowell's and MacFarlane and Kolodny's picture have serious problems, and cannot adequately treat Chisholm's puzzle.

7.8 Conclusion

I have argued for two main claims. First, there is much to learn about normative requirements from an examination of Chisholm's puzzle of ctd obligations and related issues in deontic logic. Second, given that the solution to those puzzles involves interpreting conditional obligations as in dyadic deontic logic, so too the correct interpretation of normative requirements is as in dyadic deontic logic. I hasten to add that I take the argument for the first claim to be stronger than the argument for the second one. Maybe I'm wrong and dyadic deontic logic does not provide the solution to Chisholm's puzzle. Even so, the structural analogies between Chisholm's puzzle and the issues that arise for normative requirements are so strong and clear that it would be surprising indeed if the correct treatment of the one did not illuminate the others.

In previous chapters I developed my theory of rational credences, and in this one I tackled the issue of what to say about beliefs and actions based on unjustified doxastic attitudes. The remaining chapters of this book are dedicated to the problem of easy rationality (chapter 8), an examination of how my theory is related to both Evidentialism and Reliabilism (chapter 9), and concluding remarks (chapter 10).

[19] Dowell thinks that the conclusion (in our case, "You ought not to tell your spouse") does not express a normative requirement (in the intuitive sense of "normative requirement," not the technical sense explored in this chapter)—see Bronfman and Dowell (2018).

[20] I illustrate with MacFarlane and Kolodny's semantics. Consider, for instance, the following model: $w = C$, $i = \{A, B, C, D\}$. "t" is true at A and B and false at C and D. "s" is true at A and C and false at B and D. Finally, $d(\{A, B, C, D\}) = \{A\}, d(\{A, B\}) = \{A\}, d(\{C, D\}) = \{D\}$. All the following four sentences come out true in MacFarlane and Kolodny's semantics for this model (where '$[If\, \phi]\psi$' is the conditional interpreted à la Kratzer): $O(t)$, $[If\, t]O(s)$, $[If\, \neg t]O(\neg s)$, $\neg t$.

[21] If the truth of $\neg t$ is presupposed in i, then $O(\neg t)$ is true in $<w, i>$, and so $O(t)$ cannot be true in $<w, i>$.

8
The Problem of Easy Rationality

8.1 Introduction

In chapter 4 I argued against Factualism and knowledge-based decision theory roughly along the following lines. Assume that inferential knowledge is possible, and in particular that knowledge can be gained from ampliative inferences—i.e., knowledge that *P* can be gained by inferring it from *Q*, where *Q* is a proposition or set of propositions such that *Q* does not entail *P*. In that case, $Cr_u(P|Q) < 1$. And yet, because *P* is known, according to Factualism the rational credence in *P* is 1. This is irrational overconfidence. Coupled with knowledge-based decision theory, this irrational overconfidence leads to irrational actions. As I said in chapter 4, this objection can be taken care of by moving from E = K to E = basic K, and from knowledge-based decision theory to basic-knowledge-based decision theory. I then went on to develop, in chapters 4 and 5, other arguments against these modified theories. But in this chapter I want to come back to that first argument against Factualism and knowledge-based decision theory.

A version of the problem that I raised for Factualism and knowledge-based decision theory afflicts a much wider range of theories. The more general problem has its historical roots in objections to Reliabilism from Fumerton (1996) and Vogel (2000). It was generalized by Cohen (2002) as an objection to any theory that admits of what he called "basic knowledge." Pryor (2013a) then noticed that a similar worry applied to an even larger set of theories, those that admitted what he termed "non-quotidian underminers," itself a generalization of Pollock's notion of undercutting defeaters. I argued against the Pollockian notion of undercutting defeaters in chapter 6, but I agree with Pryor that the problem is very general. In fact, it affects any theory that admits the possibility of knowledge and justification by ampliative inference. Call this the problem of easy rationality. After presenting the problem of easy rationality, I will suggest that the solution lies in denying the possibility of knowledge or even rational belief by ampliative inference. To be more precise, I will argue that whenever a subject has a rational credence of *n* in a proposition *P*, and this credence is inferentially justified (i.e., *P* is not part of the subject's evidence), then the subject has a non-inferentially justified credence of at least *n* in a proposition that entails *P*, and the subject's justification for assigning this credence is independent of his justification for assigning *n* to *P*.

Now, if the solution to the problem of easy rationality is that there is no knowledge or justification by ampliative inference, then aren't Factualism

Being Rational and Being Right. Juan Comesaña, Oxford University Press (2020). © Juan Comesaña.
DOI: 10.1093/oso/9780198847717.001.0001

and knowledge-based decision theory in the clear (as far as the problem of inferential knowledge goes)? If the problem for those theories assumes that there is knowledge and justification by ampliative inference, but the solution to the problem of easy rationality teaches us that there isn't, then it seems that proponents of Factualism and knowledge-based decision theory have nothing to worry about. But they do. Even if we grant my solution to the problem of easy rationality, there will be cases where inferential knowledge of a proposition *P* is gained such that the rational credence in *P* is less than 1. Thus, Factualism is still subject to the objection that it advises irrational overconfidence, and knowledge-based decision theory is still subject to the objection that it advises irrational courses of action. Moreover, granting my solution to the problem of easy rationality entails granting that there can be false rational beliefs, and so this solution is in fact incompatible with those versions of Factualism and knowledge-based decision theory that deny this.

Finally, I also consider how the problem of easy rationality applies to my own theory, and how exactly to apply my proposed solution to it.

8.2 Brief History

8.2.1 Fumerton and Vogel against Reliabilism

According to Reliabilism, a necessary condition for gaining knowledge from a source *K* is that *K* be a reliable source of beliefs—but this need not be something that the subject in question knows or even believes. Thus, vision must be a reliable source of beliefs for a subject to gain knowledge by seeing, but the subject herself need not even be aware that her visually based beliefs are visually based, let alone that her vision is reliable. Suppose then that a subject, Abigail, believes that the wall in front of her is red, and that, although Abigail doesn't initially have any meta-beliefs about her first-order belief about the wall, that belief is visually based and her vision is a reliable source of beliefs. Suppose now that sometime later Abigail reflects on the fact that her belief that the wall is red is based on vision. It seems that Abigail can now combine the two pieces of knowledge that she has: that the wall is red and that her belief that the wall is red is based on vision. A similar result can be applied to Reliabilism about justification. Reliabilism about justification entails that, as long as Abigail's visual system is reliable, the beliefs it is responsible for will be rational—for instance, Abigail will have a rational belief that the table is red, another rational belief that the wall is white, etc. Reliabilism also entails that her beliefs about where her beliefs come from will be rational, for the kind of introspection at work in these cases is also reliable. So Abigail will have a rational belief that her belief that the table is red was formed by her visual system, another rational belief that her belief that the wall is white was formed by her visual system, etc. For each one of these cases, Abigail can combine the two beliefs and conclude that her visual system worked correctly that time. Now, Abigail is rational in believing, it seems, that each time that her visual system issues in a belief it gets it right. The best explanation for this surely is that her visual system is reliable. There are other explanations, of course, such as conspiracy theories regarding benevolent demons, but these alternative explanations seem inferior to the reliability explanation. Therefore, Abigail can conclude that her visual system is reliable.

Something remarkable has happened here, Fumerton and Vogel argue. Reliabilism has it that I do not need to know, or even rationally believe, that my visual system is reliable in order for the beliefs it outputs to be rational. But the Reliabilist need not be so coy about knowledge of the reliability of my faculties, for if it is right then it is exceedingly easy to know, or at any rate to rationally believe, that my faculties are reliable: just combine the rational beliefs that are the outputs of those faculties with rational beliefs about the origins of those beliefs and voilà—you have a rational belief that your faculties are reliable. The problem, of course, is that knowledge of the reliability of your own faculties cannot really be acquired in this way—it is an illegitimate kind of "bootstrapping," as Vogel called it.

8.2.2 Cohen on Easy Knowledge

Stew Cohen argued that the problem that Vogel and Fumerton identified for Reliabilism actually applies to a wider class of theories—indeed, to any theory that denies what Cohen (2002) called the KR principle:

KR: A potential knowledge source *K* can yield knowledge for *S* only if *S* knows that *K* is reliable.

Any theory that denies KR admits that a source *K* can yield knowledge, and therefore rational belief, for *S* even if *S* does not know, or even has a rational belief, that *K* is reliable. Reliabilist theories, of course, do deny *KR*, but so do many other theories as well. For instance, Chisholm's epistemology, Pollock's direct realism, Pryor's dogmatism, Sosa's virtue theory, and Huemer's phenomenal conservatism all deny *KR*, because they all admit, for instance, that perception can yield knowledge even if the subject does not *antecedently* know that perception is reliable.[1] The "antecedently" qualifier is important because some of those authors might want to allow that, once we have enough perceptual knowledge, we can then

[1] See Chisholm (1966), Pollock (1970), Pryor (2000), Sosa (1991), Huemer (2001a).

build an argument that can yield rational belief in the reliability of perception. Whether that argument is good or not,[2] Cohen's point is that it is not needed.

For take any theory that denies KR. If so, a source K can yield a rational belief that *p* for *S* even if *S* doesn't antecedently know that *K* is reliable. In addition, *S* can by introspection come to rationally believe that K is responsible for the belief that *p*. We can, as before, repeat this as many times as we want, and then conclude that K is reliable. This line of reasoning, however, is just as objectionable, Cohen argues, as the bootstrapping argument was for Reliabilism.

8.2.3 Pryor on Non-Quotidian Undermining

Recently, Pryor (2013a) has argued that the problem that Vogel and Fumerton raised for Reliabilism and Cohen generalized to any theory that rejects KR applies even more widely. For Pryor, any theory that accepts the existence of what he calls "non-quotidian undermining" is subject to a similar problem. "Underminer" is Pryor's term for "defeater." Thus, a proposition that *q* undermines the rationality of a belief that *p* by a subject *S* just in case, if *S* were to become rational in believing that *q*, then *S* would no longer be rational in believing that *p*. Notice that the rationality in question here is, for Pryor, *ex ante* rationality. Even if *S* doesn't believe either *p* or *q*, the *ex ante* rationality of his belief in *p* can be undermined by the *ex ante* rationality of his belief in *q*. A proposition that *q* is, in particular, a "non-quotidian" underminer just in case *S* didn't need to antecedently rule out that *q* in order to be justified in believing that *p*. Thus, the exogenous defeaters of chapter 6 count as non-quotidian underminers in Pryor's sense.

Although Pryor doesn't mention Reliabilism, it arguably is compatible with the existence of non-quotidian underminers. For Reliabilists will in general accept that justifiedly believing that one's perception is unreliable undermines the rationality of believing the deliverances of perception,[3] but, of course, according to Reliabilism one need not rule out the unreliability of perception in order to gain justified beliefs via perception. Pryor's own dogmatism, as well as the positions mentioned earlier defended by Chisholm, Pollock, and Huemer, also accept the possibility of non-quotidian undermining. All these views hold that the rationality of certain beliefs (like, say, the belief that the table is red) is defeated by learning that certain conditions obtain (such as the fact that the table is illuminated by red lights), even though the rationality of the belief does not depend on the subject's ruling out that the conditions obtain.

It is then clear that *some* theories that are compatible with the existence of non-quotidian undermining (such as the ones just mentioned) are susceptible

[2] See Alston (1986) for an argument against it.
[3] See Goldman (1986) for his appeal to defeaters.

to the charge that they sanction as legitimate kinds of reasoning that are clearly illegitimate.

Take now *any* theory that accepts the existence of non-quotidian undermining of the rationality of a belief that p by a belief that q. Presumably we can know, according to these theories, what they claim to be the case: that the rationality of a belief that p can be undermined by the rationality of a belief that q, even if we do not need to antecedently rule out the possibility that q in order to be rational in believing that p. But now suppose that we do come to be rational in believing that p. It should be possible for us to also come to know, at least some times, that we are rational in believing that p, without this committing us to any strong claim about iteration principles in epistemology.[4] We can then combine our belief that we are rational in believing that p with our belief that this wouldn't be so if we were rational in believing that q and conclude that it is not rational for us to believe that q.

8.2.4 Huemer on Defeasible Justification

Michael Huemer has also argued (in Huemer (2001b)) for the existence of a very similar problem for any theory that admits of what he calls defeasible justification. Huemer's presentation of the problem comes closest to my own (to be presented shortly). The main differences are that he ties defeasibility to ampliative inference in ways that I would warn against, and that he thinks the solution is to give up even very plausible forms of the closure principle (more on this below).

I turn now to my own version of the problem, which will be formulated for the case of credences.

8.3 The Problem

The problem of easy rationality in its most general form arises for any theory that accepts four very plausible principles.[5] To be more precise, it arises for any theory that accepts that what those principles together imply is true of some cases, even if the theories reject the principles in their full generality. This is important, because rejecting the details of the principles does not help as long as what they jointly say is true of some cases.

[4] But, for a defense of such principles, see Cohen and Comesaña (2013a).

[5] Sections 8.3 and 8.4 borrow from Comesaña (2013b) and Comesaña (2013a). Cf. Pryor (2013c) and Pryor (2013b).

8.3.1 Ampliativity

The first principle has to do with the idea that knowledge through ampliative inference is possible:

Ampliativity: It is possible for a subject S to be inferentially justified in believing *H* on the basis of some evidence E even if *S* does not have independent justification for believing a proposition *P* such that *P* and E together entail *H*.

The qualifier that *S* need not have *independent* justification for believing *P* will prove important to our discussion later on. Ampliativity is an interesting principle. On the one hand, it might strike many as obviously true. For it might seem as if it underlies the very possibility of, say, gaining knowledge through induction. In the best case, I know the premises of a good inductive argument, and on that basis I come to know its conclusion. I need not know anything that entails the conclusion, and I need not have the same degree of confidence I have in the conclusion in anything that, together with the original premises, entails the conclusion. It might seem as if denying Ampliativity leads to a very general kind of skepticism. On the other hand, as we shall see, Ampliativity is incompatible with other very plausible principles. I will argue that we should reject Ampliativity. The resulting view does not represent a capitulation to wholesale skepticism.

8.3.2 Single-Premise Closure

The second principle is a closure principle. Principles of closure have received a lot of attention in recent decades. A useful distinction is that between Multi-Premise Closure principles and Single-Premise Closure principles. One formulation of a Multi-Premise Closure principle is the following:

Multi-Premise Closure: If *S* is justified in believing each member of a set of propositions *E* and *E* entails *P*, then S is justified in believing *P*.

The principle is formulated for *ex ante* justification. There are several issues with such a Multi-Premise Closure principle that have been discussed. One set of issues has to do with the fact that mere entailment from rationally believable propositions doesn't seem to suffice for rationality. If the subject is completely unaware of the entailment, the fact of the entailment by itself doesn't seem to have any normative significance. This is related to the problems of logical omniscience discussed in chapter 2, and I will not further discuss those issues now, because in the applications of closure that we are interested in, the entailments in question are known by the subject.

Another set of issues with Multi-Premise Closure is germane to our discussion, and it has to do with the accumulation of epistemic risk. A familiar theorem of the probability calculus is that the probability of a conjunction can never be higher than, and it will in the usual case be lower than, the probability of its least probable conjunct. If so, it may well be that even though we are rational in believing each member of a set of propositions, we are not rational in believing their conjunction. The epistemic risk associated with each proposition accumulates over the conjunction, and as it goes up it may well get high enough that we no longer count as being rational in believing the conjunction. Moreover, one need not buy wholesale into probabilism to see that this might generate counterexamples to Multi-Premise Closure.

For this reason, many philosophers are weary of accepting Multi-Premise Closure, even bracketing issues having to do with whether the entailment is known or not. A principle of Single-Premise Closure might seem to avoid this problem:

Single-Premise Closure: If *S* is justified in believing a proposition *P*, and *P* entails *Q*, then *S* is justified in believing *Q*.

There is a theorem in the probability calculus that is closely associated with Single-Premise Closure (if *P* entails *Q*, then $P(Q) \geq P(P)$), but a probabilistic understanding of justification is not assumed here. As with Multi-Premise Closure, Single-Premise Closure must also deal with the aforementioned issue of whether the entailment should be known. But it seems to avoid the problem of the accumulation of epistemic risk because there is nothing to accumulate: we are talking about a single proposition *P* which entails another proposition *Q*. There is no conjunction to worry about.

Some authors have argued, however, that even Single-Premise Closure must deal with a problem of accumulation of epistemic risk. In particular Maria Lasonen-Aarnio, Joshua Schechter, and Jim Pryor have argued against even Single-Premise Closure principles along the following lines.[6] Suppose that Mary is a mathematician who competently deduces[7] a certain mathematical proposition *Q* from a proposition *P* she knows to be true, thereby coming to know *Q*. The corresponding proof is long and complicated, but Mary is a skilled mathematician, and she performs the deduction flawlessly. In one continuation of the story, many of Mary's colleagues tell her that there is a subtle flaw in the proof. Mary doesn't see the alleged flaw, but she rationally defers to her many colleagues and lowers

[6] See Lasonen-Aarnio (2008), Schechter (2013), and Pryor (2018). Lasonen-Aarnio discusses the closure of knowledge, not rationality.

[7] The "competent deduction" formulation comes from Hawthorne (2004b), who in turn takes it from Williamson (2000).

her confidence in *Q*. In another, perhaps more realistic, continuation of the story, Mary reflects on the fact that the proof is complicated and that she is an excellent but fallible mathematician, and on this basis alone lowers somewhat her confidence in *Q* (enough to no longer count as full-out believing *Q*). In either of those cases, even Single-Premise Closure principles are violated—Mary knows that *P* and competently deduces that *Q* from *P*, and yet fails to be rational in believing *Q*.

I will argue later that this Lasonen-Aarnio/Schechter/Pryor attack on Single-Premise Closure, far from militating against my use of Single-Premise Closure, supports it. I move on to the remaining two principles.

8.3.3 Mere Lemmas

I call my next principle the "Mere Lemmas" principle, but the name deserves some explanation. The explanation will also serve as a justification of the principle.

Lemmas in mathematics or logic may well be heuristically indispensable, in the sense that we would just not see why *Q* follows from *P* unless we could see that there is a proposition *R* that follows from *P* and from which *Q* follows. The indispensability of lemmas is related to the problem of logical omniscience. But no lemma is epistemically indispensable. What I mean by saying that no lemma is epistemically indispensable is that it cannot happen that my evidence justifies me in assigning credence *n* to *P*, and yet I'm justified in assigning a different credence *m* to *P* on the basis of some lemma inferred from my evidence. Evidentially justified credences are justified by your evidence alone. The credence that you assign to any other proposition "between" your evidence and *P* is irrelevant to the credence you assign to *P*. Although you may need to notice that *P* entails *R* and that *R* entails *Q* in order to "see" that *P* entails *Q* (and, thus, in order to be justified in believing *Q*), it is *P*, and not *R*, that justifies you in believing *Q*.

Suppose, for instance, that you start out by knowing that I have a pet, but you don't know what kind of pet it is.[8] Then you come to know that it is a hairless pet. This justifies you in assigning a certain credence to the proposition that my pet is a hairless dog. Indeed, that my pet is hairless is naturally taken as incremental evidence for the proposition that my pet is a hairless dog, in the sense that the conditional credence in my having a hairless dog given that my pet is hairless is higher than its unconditional credence.[9] Of course, that my pet is a hairless dog entails that my pet is a dog, and this is not a particularly hard entailment to see. But the credence you are justified in having in the proposition that my pet is a dog

[8] I take the example from Pryor (2004).

[9] I talk here and in the formulation of the Mere Lemmas principle of conditional credences for the sake of ease of exposition, but no commitment to Probabilism should be read in this use. Rather, when I talk in this context of the conditional credence of *P* given *Q* I mean no more than the credence that is justified in *P* when one's basic evidence is *Q*.

is its credence conditional on your basic evidence, not some higher credence, and in particular not your credence conditional on the proposition that I have a hairless dog (that credence is 1!). Moreover, the fact that you are now justified in assigning a higher credence to the proposition that my pet is a hairless dog does not mean that you are now justified in assigning a higher credence to the proposition that my pet is a dog. For the rational credence that my pet is a dog conditional on my having a hairless pet is arguably *lower* than its unconditional credence.[10]

In this case, that my pet is a hairless dog is a mere lemma, and thus cannot be epistemically indispensable. That proposition does not have justificatory powers of its own. You are rational in assigning it only whatever credence it has conditional on your basic evidence, and so you should not, in particular, conditionalize on it. The Mere Lemmas principle codifies this idea:[11]

Mere Lemmas: If *Q* is inferentially justified by *P* for *S*, then *Q* can justify *S* in believing *R* only if *P* justifies *S* in believing *R*.

Thus, to come back to the example from Pryor, that my pet is a hairless dog does not justify you in believing that my pet is a dog because your basic evidence (that my pet is hairless) does not justify you in believing that proposition.

8.3.4 Entailment

Finally, suppose that *P* entails *Q*. Could *Q* then justify you in rejecting *P*? Initially, at least, it seems as if it could not. After all, if *P* entails *Q*, then if *Q* is true things are as you would expect them to be if *P* were true—how could that be a reason to reject *P*? The following principle is meant to capture that idea:

Entailment: If *P* entails *Q*, then *Q* cannot justify *S* in disbelieving *P*.

Notice that, again, given Probabilism the principle of entailment more or less follows as a theorem.[12] But given that, as was the case with Single-Premise Closure, this can be seen as much as an indictment of Entailment as a defense of it, I will not rest my defense of this principle on its connection to Probabilism.

I will have more to say in defense of Entailment and Single-Premise Closure, but let me now show how the four principles are in conflict.

[10] This is then a counterexample to Hempel's infamous "Special Consequence Condition." For more on issues of evidential transitivity, see Comesaña and Tal (2015) and Tal and Comesaña (2015).

[11] See also Weisberg (2010), who argues for a similar principle he calls "No Feedback."

[12] More precisely, for any *P* and *Q* such that both $Cr_u(P)$ and $Cr_u(Q)$ are non-extreme, if $Cr_u(Q|P) > Cr_u(Q)$, then $Cr_u(P|Q) > Cr_u(P)$.

8.3.5 The Conflict

Assume, with Ampliativity, that a subject *S* is rational in believing *H* on the basis of *E* without having independent justification for believing any other proposition *P* such that *P* together with *E* entails *H*. Notice that *H* obviously entails $H \vee \neg E$ (which is equivalent to $E \supset H$). Therefore, by Single-Premise Closure *S* is rational in believing $H \vee \neg E$. But, of course, *E* and $(H \vee \neg E)$ entail *H*. Therefore, if *S* is justified in believing *H* on the basis of *E*, then there is a proposition that *S* is justified in believing and that together with *E* entails *H*.

Notice that this is close to, but not quite, the negation of Ampliativity. For Ampliativity denies that there will be any such proposition that *S* is *independently* justified in believing, and for all we have said *S*'s justification for believing $(H \vee \neg E)$ is not independent. Independent of what? Of *S*'s justification for believing *H* itself. For all we have said so far, *S* might be justified in believing $H \vee \neg E$ on the basis of *H*, or on the basis of *E* itself.

But, given Mere Lemmas, *H* cannot justify *S* in believing any proposition unless *E* does. Therefore, the only option left open, short of denying Ampliativity, is to argue that *E* itself justifies *S* in believing $H \vee \neg E$. But that is incompatible with Entailment. For notice that for *E* to justify *S* in believing $H \vee \neg E$ is for *E* to justify *S* in disbelieving its negation, i.e., $E \wedge \neg H$. But, of course, $E \wedge \neg H$ entails *E*, and so Entailment has it that *E* cannot justify *S* in disbelieving $E \wedge \neg H$—i.e., *E* cannot justify *S* in believing $H \vee \neg E$.

I call this the problem of easy rationality because we end up being justified in believing an empirical proposition $(H \vee \neg E)$, but not on the basis of any evidence for it. As we shall see, my own preferred solution to the problem is to embrace this consequence. But first I consider other possible resolutions, ones that deem the consequence sufficiently repellent to warrant rejecting it.

8.4 Possible Solutions

8.4.1 Denying Single-Premise Closure

As I said, this is the solution preferred by Huemer (2001b) to a similar problem. Others have also urged that we should give up Single-Premise Closure.[13] I will not canvass all possible reasons for denying closure, but I do want to come back to the Lasonen-Aarnio/Schechter/Pryor line of reasoning anticipated above.

Recall that that line of reasoning had it that the same misgivings we should have about multi-premise closure related to the accumulation of epistemic risk across

[13] See, for instance, Sharon and Spectre (2017), and cf. Comesaña (2017).

multiple premises we should also have about the accumulation of epistemic risk across complicated deductive inferences, even if they are single-premise.

Now, one fact that we should keep in mind is that the applications of Single-Premise Closure that lead to the problem of easy rationality are not particularly hard ones: it is just a single application of disjunction introduction. Moreover, we are guaranteed to have the conceptual resources to comprehend the two propositions we are adding, for one is just the proposition which we are inferentially justified in believing (*ex hypothesi*), and the other one is just the negation of the evidence on the basis of which we are supposed to be so justified. So, to the extent that the attack on Single-Premise Closure relies on the difficulty or length of the deduction in question, it doesn't affect the use to which we are putting Single-Premise Closure.

But, in addition, far from being an argument against my use of Single-Premise Closure, the line of attack in question supports it. There are two relevant arguments here: one from E to H and another from H to $H \vee \neg E$. If the attack on Single-Premise Closure works, then it should also work (with perhaps even more force) for non-deductive inferences, like the inference from E to H. Translated to this case, what Lasonen-Aarnio, Schechter, and Pryor are saying is that your confidence in H cannot be higher than your confidence in the proposition that $E \wedge \neg H$ is false. But notice that the proposition that I say you must be justified in believing, if you are justified in believing H on the basis of E, is the disjunction $H \vee \neg E$. Reasons for raising one's confidence in $E \wedge \neg H$ will be reasons for lowering one's confidence in $H \vee \neg E$. So, Lasonen-Aarnio, Pryor, and Schechter are in effect arguing that E can justify you in believing H only to the extent that it justifies you in believing $H \vee \neg E$. Therefore, their attacks on Single-Premise Closure actually support the use to which we are putting the principle here.

In line with my first comment regarding the option of denying Single-Premise Closure, it is well worth remembering that reasons for doubting that the principle holds in full generality are not automatically reasons for thinking that it fails precisely in those cases that generate the problem of easy rationality. I have not seen any reason for thinking that this is the case.

8.4.2 Denying Mere Lemmas

Klein (1995) has in effect denied Mere Lemmas. To be completely fair to Klein, he has defended the conditional claim that if we are to hold on to plausible closure principles, then Mere Lemmas has to go (although I think that Klein is clear that he does want to hold on to those closure principles, and so the stronger claim is also true). This also strikes me as a non-starter, though I'm afraid that, in this case, I do not have much to say to someone convinced that Mere Lemmas is false. To deny Mere Lemmas is to attribute magic powers to mere lemmas; it is to accept that epistemic justification can be created *ex nihilo*; more importantly, it is just to get the cases wrong.

8.4.3 Denying Entailment

Some philosophers deny Entailment.[14] As I said, Entailment can be supported by probabilistic considerations. For it is a theorem of the probability calculus that if H entails E then $P(H|E) \geq P(H)$—at least when $P(E) < 1$ and $P(H) > 0$. Reasons to be wary of Probabilism, however, may well be reasons to be wary of Entailment, or will at least deny Entailment this line of support. Of course, this need not be the case with *every* reason to be wary of probabilism. In particular, misgivings having to do with our rational lack of logical omniscience do not seem to translate to doubts about Entailment, or at least to the use to which we put Entailment when deriving the problem of easy rationality. For, to repeat what has by now become a refrain of this chapter, the particular use to which we put Entailment does not seem to rely on any problematic assumption of logical omniscience—we are just pointing out that $E \wedge \neg H$ entails E, and so the latter cannot be a reason to reject the former. The entailment in question is a benign application of conjunction elimination, hardly meriting complains about logical omniscience.

Now, it may be pointed out that even if the particular instances of conjunction elimination to which are appealing do not themselves merit charges of assuming logical omniscience, the justification of Entailment itself relies on the fact that it follows from the axioms of probability. But the derivation of that result is itself not particularly hard. Be that as it may, I take the point that we should not completely rely on a probabilistic defense of Entailment. I do not think we have to. For consider this other principle, a version of which Carolina Sartorio and myself have defended on independent grounds (see Comesaña and Sartorio (2014)):

Epistemic difference-making: E justifies S in believing H only if $\neg E$ does not justify S in believing H.

The epistemic difference-making principle itself follows from a probabilistic understanding of evidential support, but one need not buy into such an understanding to adopt it. To begin with, it is intrinsically plausible. In addition, it follows from plausible versions of reflection principles. Suppose that H is an empirical proposition, and that you are not presently justified in either believing or disbelieving H. You do want to find out whether H, and so you plan on performing an experiment which will help answer whether H is true. The

[14] See, for instance, Pryor (2013b) and Vogel (2014).

experiment will tell you tomorrow whether E or $\neg E$ is true. So, you know now that tomorrow you will find out whether E or $\neg E$, and you also know now that no epistemically bad things will happen between now and tomorrow (you will not forget or in any other way lose any evidence you now have, you will not suddenly become irrational, etc). Suppose now that the epistemic difference-making principle fails in this case and both E and $\neg E$ would justify you in believing H. This may already strike you as absurd (it does so strike me). But things get worse. Because it seems that you should not wait and start believing H right now. For you know that, no matter what, tomorrow you will be justified in believing H, and you know that the only epistemically relevant difference between your situation now and tomorrow is that tomorrow you will have strictly more evidence than you do now. But this is of course absurd. By hypothesis, you do not *now* have any evidence for believing H, and, we can suppose, you can only be inferentially justified in believing H. The obvious culprit in this situation is the denial of the epistemic difference-making principle.

The epistemic difference-making principle entails Entailment. For suppose that H entails E (this is the antecedent of Entailment). In that case, of course, $\neg E$ entails $\neg H$. Assuming that the entailment in question is sufficiently obvious to the subject (an assumption that in any case we are making throughout), $\neg E$ justifies S in believing $\neg H$. Given the epistemic difference-making principle, E cannot justify S in believing $\neg H$ as well, which is the consequent Entailment.

For these reasons, I also take Entailment (or, to be more precise, the uses to which I am putting the principle) to be non-negotiable.

8.4.4 Denying Ampliativity

The only possibility left open is, of course, to deny Ampliativity. That is what I want to do, and I want to explain why doing so is not as bad as it may seem.

Notice, first, exactly why we are denying Ampliativity. Whenever a subject is justified in believing a proposition H on the basis of what we might want to call an inductive argument for H with some (plausibly conjunctive) premise E, that subject will also be independently justified in believing the associated conditional of that argument (which is equivalent to the proposition $H \vee \neg E$). This doesn't mean that the inference the subject performs must "go through" that conditional. The subject's *ex post* justification for believing the conclusion need not be based in the credence she assigns to the conditional, even if she will have *ex ante* independent justification for believing the conditional. Therefore, denying Ampliativity is perfectly compatible with the existence on inductive arguments in this sense.

We saw in section 8.3.5 that neither E nor H can justify me in believing that disjunction, so a natural question to ask is: what *does* justify me in believing it? At this point, the answer to that question should not be surprising. We have seen that, in the Bayesian framework, basic epistemic principles are encoded in conditional probabilities, and these conditional probabilities in turn constrain unconditional probabilities. In particular, we saw in chapter 4 that $Cr_u(H|E) \leq Cr_u(H \vee \neg E)$. If we are Subjective Bayesians this means that whenever a subject adopts an epistemic principle according to which E is good evidence for H, she will thereby adopt some highish credence in $H \vee \neg E$. If we are Objective Bayesians, this means that, if E is objectively good evidence for H, then we are a priori justified in assigning some highish credence to $H \vee \neg E$. And even if we are not any kind of Bayesian because we take seriously the problem of logical omniscience, that is no obstacle to accepting the consequence that there is a correlation between which epistemic principles are correct and which propositions we are a priori justified in believing. This is a story that meshes well with the solutions to problems similar to the problem of easy rationality adopted by Cohen (2010) and Wedgwood (2013). Denying Ampliativity may seem to commit one to wholesale skepticism and to mysterious sources of justification. But it doesn't, insofar as no measure whatever of skepticism is entailed by the denial of Ampliativity, and the mystery of a priori justification is one that we have to learn to live with anyway.

8.4.5 The Problem of *Ex Post* Easy Rationality

Notice that there are two problems of easy rationality: an *ex ante* one and an *ex post* one. We have been looking so far at the *ex ante* problem: Single-Premise Closure entails that whenever some evidence E provides me with *ex ante* inferential justification to believe some proposition H, I also have *ex ante* justification to believe $H \vee \neg E$. Meanwhile, Mere Lemmas entails that H cannot provide me with such justification, and Entailment that E cannot either. The solution is to deny Ampliativity: we have *ex ante* justification for believing $H \vee \neg E$ which is independent of both H and E. But what if a subject simply ignores this independent justification and infers $H \vee \neg E$ from H or E? The question is not now one about the *ex ante* justification the subject has for believing the disjunction—let us suppose we agree that there is independent *ex ante* justification that is always available. The question is rather about the *ex post* status of the subject's belief in the disjunction, which simply ignores that *ex ante* justification. Is it justified?

My answer will be that it isn't. But notice that an *ex post* version of Single-Premise Closure entails that it is. To a rough approximation (the details won't matter), an *ex post* version of Single-Premise Closure has it that if a subject has *ex post* justification for believing a proposition H and believes a proposition P on the basis of the fact that H entails P, then the subject is *ex post* justified in believing P on that basis. Readers may be familiar with *ex post* versions of Single-Premise Closure under the name of *transmission* principles, where the idea is that the subject's justification for believing H is transmitted to P.

But now, recall Mere Lemmas and Entailment. Mere Lemmas has it that H can justify any P which it entails only if whatever justifies H justifies P. When the P in question is $H \vee \neg E$, Entailment entails that E itself cannot justify it. Therefore, transmission principles are incompatible with Mere Lemmas and Entailment.

Some may, of course, take this as a reason for rejecting either Mere Lemmas or Entailment. After all, isn't deduction a paradigmatic way of extending the scope of one's justified beliefs?[15] Of course it is, in general. But epistemic dependence relations do not always mirror entailment relations. Indeed, I take it that the right moral to draw from alleged counterexamples to closure such as Dretske's painted mule case (see Dretske (1970)) is not that Single-Premise Closure is false, but rather that transmission principles are false. That the animal in the pen is a zebra entails that it is not a painted mule, but our justification for believing that it is a zebra depends on our justification for believing that it is not a painted mule, not the other way around. Therefore, the incompatibility of Mere Lemmas and Entailment with transmission principles does not give us a reason to abandon either of the first two, because we have reason to reject transmission principles. If, in the cases we have been considering, a subject believes $H \vee \neg E$ on the basis of either E or H, then she is not justified.

8.5 Epistemic Principles and Evidence

Does denying Ampliativity offer solace to proponents of Factualism in the face of one of my arguments in chapter 4 against them? No, it doesn't. Recall that the argument in question charged Factualism with generating irrational overconfidence in propositions that are inferentially known. Denying Ampliativity does not help Factualism with that problem. Denying Ampliativity tells us that whenever a subject is justified in assigning credence n to a proposition H, she will also be justified in assigning credence of at least n to a proposition that, together with the original premise E, entails H. But the n in question need not be 1. If we want to say that the subject may still know H under those conditions, there is still a problem for Factualism. For then H will be part of the subject's evidence, and this entails the kind of irrational overconfidence at the center of the argument.

There are additional reasons for thinking that the disjunctions corresponding to fundamental epistemic principles are not part of the subject's evidence. For, whenever E is part of the subject's evidence, counting $H \vee \neg E$ as also part of her evidence would exaggerate her epistemic position with respect to H. If both E and $H \vee \neg E$ are part of the subject's evidence, then she need not consider states where H is false in figuring out what to do (as well as what to believe). Therefore, even

[15] Compare Williamson (2000).

when the subject is, as I argued that she must be, justified in believing the disjunction, that disjunction should not be counted as part of her evidence even when true.[16]

That those disjunctions should not be counted as part of the subject's evidence fits with two themes of this book. The first one is that inferentially acquired beliefs are not part of our evidence. This is the rationale behind the move from E = K to E = basic K, and from E = R to E = basic R. Although those disjunctions are not strictly speaking evidentially acquired, they are based, at least indirectly, on epistemic principles. It is because E justifies H that we are justified in believing $H \vee \neg E$. Indeed, the epistemic principle embodied in $Cr_u(H|E)$ justifies us in assigning a credence of at least $Cr_u(H|E)$ in $H \vee \neg E$. The justification we have for believing those disjunctions is not evidential: it is not based on some other proposition we are justified in believing. But it is also not like the basic justification we have for believing the content of our undefeated experiences. It is, if you want, inferential but non-evidential justification.

The second, related, reason for thinking that those disjunctions should not be part of our basic evidence is that they do not have their origin in experience. I am dealing in this book with the justification for our empirical beliefs. A broadly empiricist approach permeates the whole enterprise. Subjects deprived of all sensory input do not have any basic empirical evidence. If our justification for believing the disjunctions in question depends entirely on the fact that E evidentially supports H, then those disjunctions cannot be part of our basic empirical evidence: they just don't have the right provenance in experience.

Notice that this justification for excluding the conditionals in question from our basic evidence is not *ad hoc*, insofar as it applies to other propositions for which we also have some measure of justification. We have seen before, for instance, that the interdependence of conditional and unconditional probabilities has the consequence that we must favor some hypotheses over others even in the absence of any evidence underlying such asymmetric treatment. For instance, even without having observed any emeralds we should be more confident that all emeralds are green than we are that all emeralds are grue. But, of course, it is not part of our basic empirical evidence that all emeralds are green. There is therefore precedent to exclude propositions like those disjunctions from the stock of our basic evidence.

[16] Some might be tempted to argue that the subject is not even justified in believing the disjunctions, perhaps on contextualist or subject-sensitive grounds. But notice that we do want to say that the assumption is that the subject can be justified in believing H. Given that she will have the same degree of justification for believing the disjunction, that H entails the disjunction, and that the relevant practical context stays fixed, it is hard to see how these threshold-shifting maneuvers can get a hold here. Of course, another option is to say that the subject is not really justified in believing H, but only in assigning it a high credence.

8.6 The Problem of Easy Rationality for Experientialism

There are two ways in which the problem of easy rationality applies to the theory developed in this book. The obvious one is that my theory too admits of ampliative inference—at least, I haven't said anything which implies that it doesn't. Suppose, for instance, that it is part of my evidence that sufficiently many green emeralds have been observed. I can, on this basis, infer that all emeralds are green, or at least come to have a sufficiently high credence in that proposition. But then, that proposition entails that either not sufficiently many green emeralds have been observed or all emeralds are green. Therefore, by Single-Premise Closure I am justified in assigning this proposition the same credence I assign to the proposition that all emeralds are green. But what can justify me in assigning that credence? Not that all emeralds are green, by Mere Lemmas, nor that sufficiently many green emeralds have been observed, by Entailment. We know by now the solution to this problem: I have independent justification for believing the conditional in question, and so Ampliativity is false.

That is how the problem arises, and how to solve it, for a case that the theory itself treats as evidential. According to Experientialism, however, the relationship between an experience and belief in its content is not evidential. The experience provides its content as evidence, but it is not itself evidence for that belief. Doesn't the problem of easy rationality arise also for these basic beliefs, however? Suppose, for instance, that my experience has the representational content that the wall in front of me is red. According to Experientialism, that experience provides me with the proposition that the wall in front of me is red as evidence. This means that I am justified in believing that the wall in front of me is red. Now, by Single-Premise Closure we conclude that I am also justified in believing the material conditional that if my experience represents the wall as red, then the wall is red. In the evidential case, Mere Lemmas entails that a hypothesis H can justify me in believing $E \supset \mathrm{H}$, where E is my evidence, only if E itself so justifies me, because H is itself justified by E and doesn't have justificatory powers of its own. Entailment, in turn, entails that E itself cannot justify me in believing $E \supset \mathrm{H}$, and so I concluded that we should reject Ampliativity and hold that I have justification for believing that conditional which is independent of both E and H. But, given that according to Experientialism the relationship between the experience and its content is not evidential, we cannot apply the same reasoning here. That the wall in front of me is red is non-evidentially justified for me, and so it is not a Mere Lemma in the required sense. Still, doesn't it seem obviously bad to say that just by looking at the wall I can conclude that my experience correctly represented the color of the wall this time?

It is important to notice, first of all, that all that follows from the fact that I am justified in believing that the wall in front of me is red is that it is not the case that my experience represents the wall as red and the wall isn't red. This does not amount to saying that my experience was veridical, for it is compatible with my not having had any experience with the content that there is a red wall in front of me to begin with. But it might be held that I can come to know by introspection that I have an experience with a certain content whenever I do. If so, we can combine this introspective knowledge with knowledge of the material conditional to conclude that my experience was veridical. This move raises questions about the powers of introspection, but in what follows I grant that something like this works.

The question was, then, isn't it bad to say that just by looking at the wall I can conclude that my experience correctly represented its color? And the answer is that yes, that consequence is bad, and for roughly the same reason that it is bad for the proponent of Psychologism (according to which the experience is evidence for a belief in its content). True, according to Psychologism the relationship between the experience and a belief in its content is that of evidence to a proposition justified by that evidence, whereas according to Experientialism that relationship is that of provider of evidence to evidence thereby provided. But there is still an epistemic dependence between the experience and a belief in its content according to both theories. According to both theories, for instance, a belief in the content of an experience would not (in the ordinary case) be justified in the absence of such an experience.[17] Given that there is an epistemic dependence between evidence and its possession conditions, an analogue of Mere Lemmas holds for the relationship between them as well:

Mere Lemmas (evidential possession): If an experience with the content that *P* justifies *S* in believing *P* (by providing *P* as evidence for *S*), then *P* can justify *S* in believing *Q* only if the fact that *S* has an experience with the content that *P* justifies *S* in believing *Q*.

And the same goes for an analogue of Entailment:

Entailment (evidential possession): If a proposition *P* entails that *S* has an experience with the content that *Q*, then having an experience with the content that *Q* cannot justify *S* in disbelieving *P*.

Notice that both Psychologism and Experientialism agree that having an experience with a certain content justifies subjects in believing that content. The difference, once again, is that whereas proponents of Psychologism think that experiences justify in virtue of being evidence for their contents, Experientialists think that they justify in virtue of providing that content as evidence. Now, if it is false that *S* has an experience with the content that there is a red wall in front of

[17] I add the parenthetical "in the ordinary case" because justification for a belief in the content of an experience can be overdetermined, and so still be justified even in the absence of the experience.

him, this fact can justify him in believing the disjunction that either he doesn't have an experience with the content that there is a red wall in front of him or there is a red wall in front of him. By an application of the evidential possession version of Entailment, then, that *S* does have an experience with the content that there is a red wall in front of him cannot justify him in believing the same proposition. The evidential possession version of Mere Lemmas then entails that, according to Experientialism, an experience with the content that there is a red wall in front of him cannot justify a subject in believing that that experience is veridical.

A related worry has to do with rationality-based decision theory. According to that view, the states that we should consider when deciding what to do are exactly those that are compatible with our evidence, where the theory of evidence is provided by Experientialism. This view has it, then, that when having an experience with the content that there is a red wall in front of you, you should conditionalize Cr_u on the proposition that there is a red wall in front of you. That means that you should assign credence 1 to that proposition. But then, you should also assign credence 1 to any proposition it entails—in particular, to the proposition that your experience did not lead you astray this time. Isn't that the same kind of irrational overconfidence that I accused Factualism of entailing?

The answer to that question is "No." My objection to Factualism was that it generates irrational overconfidence in cases of evidential knowledge. That was because Factualism has it that we should assign credence 1 to evidentially acquired knowledge. But Experientialism does not have this consequence. My objection to E = K was that not all knowledge is evidence. My objection to E = basic K was that not all evidence is knowledge. The claim of irrational overconfidence is the basis for my claim that not all knowledge is evidence, but not for the claim that not all evidence is knowledge. The argument for this last claim had to with the proliferation of practical and theoretical dilemmas and the violation of minimalist understandings of the claim that justification is action-guiding. I am not being hypocritical in criticizing opposing views for having features shared by my view.

But still, isn't it bad that my view entails that in some cases I should assign credence 1 to the proposition that my experience did not lead me astray? I don't think it is bad, for two reasons. First, the proposition I should assign credence 1 to is the proposition that it is not the case that I had an experience with the content that the wall in front of me is red while the wall in front of me is not red. We should not describe this proposition as saying that my experience was veridical this time, or even that my experience did not lead me astray, because (as I mentioned) this proposition is compatible with my not having had any experience at all, and is therefore completely silent on what connection (if any) there is between my experience and my belief. I am assuming that I do know, by introspection, that I have had an experience with the content that the wall in front of me is red, and this bit of knowledge, combined with the proposition in question, does entail that my experience did not lead me astray—but we are not forced to claim that I should assign credence 1 to this proposition unless we claim that we should assign credence 1 to my introspective knowledge.

Second, we should consider the issue of assigning credence 1 in the context of the particular version of the Bayesian framework that I am advocating. Traditional Bayesianism, with its commitment to Conditionalization as an updating rule, has it that credence 1 is forever. This is just plainly unacceptable. But the kind of Bayesianism I advocate replaces Conditionalization with ur-prior Conditionalization, and ur-prior Conditionalization does not entail that credence 1 is forever. Thus, even if it is an inevitable consequence of assigning credence 1 to the proposition that the wall in front of me is red that I have to also assign credence 1 to the proposition that it is not the case that I had experience with the content that the wall is red while the wall wasn't red, that doesn't mean that I will now be unmoved by any information that we would naturally count as evidence that that proposition is false. As the problems presented in chapter 6 show, it is hard to account for exogenous defeaters formally. But this is a problem for every theory, Psychologism and Factualism included. My approach is to take the notion of when an experience basically justifies belief in its content (and thus when the experience provides its content as evidence) as not being formally represented, but this is not to be taken as a drawback of my theory.

8.7 Conclusion

In this chapter I have argued that whenever we are inferentially justified in believing a proposition *P* on the basis of some other proposition *Q*, we have independent justification for believing a third proposition, namely the proposition that $Q \supset P$, which together with *Q* entails *P*. The importance of this result to the themes of this book is twofold. On the one hand, the result might seem to vindicate Factualism from my argument, in chapter 4, that it results in practical and theoretical irrationality by allowing evidential knowledge to count as evidence. On the other hand, it threatens to vindicate not just E = K, but also E = R, which I also reject. I argued in the previous sections that the result does not have these consequences. E = K is not vindicated, in the first place, because the credence we are justified in assigning to the conditionals in question will not, in general, be 1. But if all knowledge is evidence, then a Factualist will sometimes be forced to assign credence 1 to those propositions. Moreover, there are reasons to think that those conditionals should not be part of our evidence, reasons that are independent of the fact that we should not, in general, assign them credence 1.

In the next chapter I compare the view I have been developing in this book with two traditional epistemological views: Reliabilism and Evidentialism. My view has features in common with both of them, but also important differences.

9
Evidentialism, Reliabilism, Evidentialist Reliabilism?

9.1 Introduction

The theory of empirical justification developed in this book can be summarized as follows. There are epistemic principles which dictate which attitudes to take towards any proposition given any body of evidence. If we bracket concerns regarding logical omniscience, these principles can be seen as encoded in an ur-prior, and the primary doxastic attitudes are taken to be credences. At any given time, the evidence a subject has is constituted by the propositions the subject has basic justification for believing. In this chapter, I want to compare the resulting theory with two of the most influential theories of empirical justification of the last few decades, Reliabilism and Evidentialism, as well as with my own previous hybrid of these views, Evidentialist Reliabilism.[1] Evidentialist Reliabilism, like its two parents, is an epistemology of full belief, but other authors have extended the theory to cover credences as well.[2] In this chapter I argue that the proper extension of Evidentialist Reliabilism to the case of credences makes it coincide with the version of Objective Bayesianism defended in this book. Because this kind of theory relies on the notion of an ur-prior that determines evidential probabilities, it is in conflict with the traditional reductivist aspirations of Reliabilism. And because it holds that some propositions are justified for a subject even in the absence of evidence for them, it is in conflict with the basic tenet of Evidentialism.

9.2 Evidentialism

Evidentialism has been defined by Conee and Feldman (1985) as follows:

Evidentialism: Doxastic attitude D toward proposition p is epistemically justified for S at t if and only if having D toward p fits the evidence S has at t.[3]

[1] See Comesaña (2010a). Other papers where I tackle the issues taken up in this chapter are Comesaña (2002), Comesaña (2006), Comesaña (2009), and Comesaña (2010b).
[2] See Tang (2016b), Dunn (2015), and Pettigrew (forthcoming).
[3] Conee and Feldman (1985), p. 83.

Being Rational and Being Right. Juan Comesaña, Oxford University Press (2020). © Juan Comesaña.
DOI: 10.1093/oso/9780198847717.001.0001

Three questions need to be answered before we have a full understanding of Evidentialism: what kinds of things can *be* evidence? What is it for a subject to *have* some evidence? And what is it for a body of evidence to *fit* some doxastic attitude?

Conee and Feldman themselves have a conception of evidence and its possession according to which two subjects in the same total (non-factive) mental state cannot differ in what evidence they possess. They take this to amount to a kind of internalism. They say:

> Somewhat more precisely, internalism as we characterize it is committed to the following two theses. The first asserts the strong supervenience of epistemic justification on the mental:
>
> > S The justificatory status of a person's doxastic attitudes strongly supervenes on the person's occurrent and dispositional mental states, events, and conditions.
>
> The second thesis spells out a principal implication of S:
>
> > M If any two possible individuals are exactly alike mentally, then they are alike justificationally, e.g., the same beliefs are justified for them to the same extent.[4]

Supervenience may be too weak a notion to capture the essence of Evidentialism. Plausibly, the Mentalist Evidentialist wants more than mere supervenience: they want not just the existence of a mere co-variation, but a *constitutive* relation between justification and evidence. If it turns out, say, that justification and mental states co-vary in the requisite way only because they in turn co-vary with a third condition, the resulting view need not be particularly friendly to Evidentialism. An analogy may help bring the point home. Suppose that we define Physicalism as the thesis that every fact supervenes on physical facts. That thesis is compatible with Cartesian substance dualism, as long as the non-physical stuff exists necessarily. Maybe the supervenience thesis is interesting in its own right, but conceiving of Physicalism as compatible with substance dualism does not get the spirit of the view right. Analogously, one would have thought that Evidentialism would have to be incompatible with non-evidential facts determining epistemic justification, even when they obtain necessarily.[5]

The Mentalist conception of evidence is not the only one compatible with Evidentialism. One can combine Evidentialism with other conceptions of evidence and its possession and end up with a package of views which denies the supervenience of epistemic justification on (non-factive) mental states. For instance, if one adopts E = K, then two subjects may be in the same (non-factive) mental states and yet differ in what evidence they have (unless one holds that whenever

[4] Conee and Feldman (2001), pp. 233–4.

[5] See Comesaña (2005c).

two subjects are internal duplicates, then their respective bodies of knowledge justify the same beliefs).

It is worth spelling out in a bit more detail how these two different conceptions of evidence and its possession result in different versions of Evidentialism. Conee and Feldman's own view on evidence and its possession is in line with Psychologism, holding (roughly) that the evidence a subject has at a time is constituted by the non-factive mental states the subject is in at that time. Prominently, the experiences the subject has at a time form the bulk of the evidence the subject has. Other non-factive mental states, such as apparent memories, may also be part of the subject's evidence. In addition, the content of the justified beliefs of the subject can also be counted as part of his evidence, if in a derivative way. The basic evidence is constituted by the non-doxastic, non-factive mental states the subject is in, and the derivative evidence is constituted by the content of the beliefs that those states justify.[6] Thus, the basic evidence, according to this conception, consists of a subset of mental states, and there isn't really a useful distinction between the evidence and its possession. To have a mental state as evidence at a time is simply to be in that mental state at the time. This is why this view is a version of Psychologism. By contrast, according to both Factualism and Experientialism, there is a strict distinction between the evidence and its possession. Evidence consists of propositions, and it is possessed only to the extent that those propositions are either known (for Factualism) or non-evidentially justified (for Experientialism).

Evidentialism as explained so far is a theory of *propositional* justification—of what it is for a doxastic attitude to be justified for a subject, independently of whether the subject adopts that attitude. We also need a theory of *doxastic* justification—of what it takes for an attitude to be justifiedly adopted. It won't do just to say that an attitude is doxastically justified just in case it is propositionally justified and adopted: subjects may adopt the right attitudes for the wrong reasons. Conee and Feldman themselves propose the following theory of "well-foundedness" to add to their Evidentialism:

S's doxastic attitude D at t toward proposition p is well-founded if and only if:
- (i) having D toward p is justified for S at t; and
- (ii) S has D toward p on the basis of some body of evidence e, such that
    - (a) S has e as evidence at t;
    - (b) having D toward p fits e; and
    - (c) there is no more inclusive body of evidence e' had by S at t such that having D toward p does not fit e'.[7]

[6] Care should be taken not to attribute to such derivative evidence more justificatory power than it deserves. This was the point of the Mere Lemmas principle of chapter 8.

[7] Conee and Feldman (1985), p. 93.

Notice that Conee and Feldman are relying here on the notion of basing an attitude on a body of evidence. The Evidentialist notion of well-foundedness is thus importantly different from the Evidentialist notion of justification. To see the difference, consider either Factualism or Experientialism. According to both theories, evidence is propositional. According to Factualism a proposition is possessed as evidence just in case it is known, whereas according to Experientialism it is possessed as evidence just in case it is basically rational. Both theories are compatible with at least the letter of Evidentialism. On either view, a proposition can be justified for a subject by either being sufficiently supported by the evidence the subject has or by being part of that evidence. There will therefore be no difference in the justificatory status of any proposition for any subject without a difference in the evidence possessed by those subjects. On those theories, however, one can be justified in believing the propositions that are part of one's evidence even if one does not base those beliefs on any evidence. Indeed, barring controversial cases of higher-order evidence, justification for believing the propositions that are part of one's evidence will in general require that one *not* base those beliefs on any evidence, according both to Factualism and Experientialism. Rather, those basic beliefs will be the deliverances of different sensory modalities. When you know that there is a snowball in front of you because you see it, it is part of your evidence that there is a snowball in front of you, and you are justified in believing that there is a snowball in front of you, but your belief that there is a snowball in front of you is not based on any evidence you have (your belief is certainly not based on itself, and you may have no other relevant evidence). So, although these views are compatible with Evidentialism, they are not compatible with the well-foundedness theory of basing, for that theory requires that all of one's justified beliefs be based on evidence. I return to this important issue in section 9.5

Why the need for clause (ii)(c)? Conee and Feldman's idea here is that even if the part of the subject's evidence on which he bases his attitude does indeed fit that attitude, there may be other parts of his evidence which don't. Suppose, for instance, that I believe that Fred is older than nine, and that I base this belief on the evidence that I have that Fred has gray hair. Compatible with all that, it may also be part of my evidence that Fred suffers from a condition that may cause premature graying of the hair. Moreover, what matters is which attitude my *total* evidence justifies—for it may also be part of my evidence that Fred goes to college. In that case, my belief would be well-founded only if based on those three relevant pieces of evidence, and it wouldn't be if based only on the color of Fred's hair.

It is important to note that Evidentialism as well as the accompanying notion of well-foundedness apply to doxastic attitudes in general, and not just to beliefs. Thus, our evidence can fit disbeliefs and suspensions of judgments as well as beliefs, and they may also fit degrees of beliefs (credences).

There are, I said, three questions that an Evidentialist needs to answer, and we gave a sketch of some of the options available to the Evidentialist as far as the conception of evidence and its possession goes. The third question, and arguably the most important one, is about fit. What does it mean to say that a body of evidence "fits" a propositional attitude? There are, broadly speaking, two options here: one may be a reductivist or a non-reductivist about evidential fit. Non-reductivists about fit think that the best we can hope for is a theory that systematizes facts about what evidence fits which doxastic attitudes. Thus, for instance, a Mentalist Evidentialist can say that evidence consisting of an experience with the content that *p prima facie* fits believing that *p*. A reductivist about fit thinks that facts about fit themselves can be reduced to other facts (perhaps facts about reliability, or counterfactual dependence, etc.). Ideally, the reductivist can recover the systematization of fit from lower-level facts via a definition of fit in terms of those lower-level facts.

Evidentialism itself is neutral between these approaches. As developed by Conee and Feldman, however, it is definitely a non-reductivist approach, where the bedrock epistemic facts are particular facts about which bodies of evidence fit which attitude, and not a general account of what fit consists in.

I'll come back to this issue regarding the Evidentialist's notion of fit. Before that, however, I take a look at Reliabilism.

9.3 Reliabilism

Hinted at by Ramsey (1931) as a component of a theory of knowledge, Reliabilism as a theory of epistemic justification was proposed by Goldman (1979). According to Goldman, epistemic justification can be defined recursively as follows:

S is justified in believing that *p* at *t* if and only if:

(i) *S*'s belief that *p* at *t* results from a belief-independent cognitive belief-forming process that is unconditionally reliable; or

(ii) *S*'s belief that *p* at *t* results from a belief-dependent cognitive belief-forming process that is conditionally reliable and all the input beliefs are justified.

Goldman later adds some comments which make it clear that the notion of being justified here is that of *prima facie* justification, which can be defeated—Goldman's account of defeat is itself explained in Reliabilist terms.

The base clause in the definition is intended to cover cases of basic, non-inferential belief formation, paradigmatically cases of perceptually acquired belief. There could, of course, be cases where some of the inputs to the process are themselves beliefs that still do not count as inferential cases. For instance, I may well become justified in believing that I believe that it is raining by reflecting on my belief that it is raining, but the process in this case is not fruitfully thought of as an inferential process. Although Goldman himself does not make this point, in an inferential process it is the *content* of the input beliefs that is best thought of as being reliably connected to the content of the justified beliefs, whereas in the case just presented it is the fact that I *believe* that it is raining that justifies my second-order belief. Notice also that Reliabilism is in the first place a theory of doxastic justification, of what it is for an episode of *believing* a proposition to be justified.

There are four issues for such a Reliabilist theory of justification: the "new evil demon" problem, the blind reliability problem, the generality problem, and the question of how to measure reliability.

We've already met the new evil demon problem in chapter 5. The new evil demon problem is the objection that reliability is not necessary for justification. The blind reliability problem, first presented by BonJour (1980), is the flipside of that: the contention that reliability is not sufficient for justification. Suppose, to take the most famous of BonJour's examples, that Norman is, under certain conditions which normally obtain, a completely reliable clairvoyant about certain subject matters—for instance, about the whereabouts of the President of the United States. Norman, however, thinks that there is nothing special about him—he doesn't think that he has clairvoyant powers, and in fact he thinks that clairvoyance doesn't exist. Nevertheless, from time to time he finds himself with spontaneous beliefs about the whereabouts of the president. Now, after a while Norman should start suspecting that there is something special about these beliefs, for they will almost invariably be confirmed. Thus, Norman will slowly start to have inductive evidence of his own clairvoyance. But let us concentrate on one of the first beliefs formed in that way. Because it was reliably formed, it seems that a Reliabilist will have to say that the belief in question is justified—but, BonJour says, this is absurd. Now, the Reliabilist might now turn our attention to the fact that the definition was of *prima facie*, defeasible justification, and hold that although the belief is indeed *prima facie* justified by the reliable process that produced it, it is nevertheless defeated by Norman's belief that beliefs that "come out of nowhere" (as his own clairvoyant beliefs appear to him) are not usually true. But BonJour insists that there will be possible cases where the subject in question does not have evidence against the reliability of the belief, while still also lacking evidence as to its reliability. If these cases are possible, then the *prima facie* justification will not be defeated, but the belief will still not be justified according to BonJour.

Both the new evil demon problem and the problem of blind reliability presuppose that we can figure out whether a particular belief-forming process is reliable. The generality problem, first noticed by Goldman (1979) himself and later pushed forcefully by Conee and Feldman (1998), questions this assumption. The generality problem starts with the claim that only *types* of processes, and not their *tokens*, are straightforwardly assessable for reliability. This is so because only types of processes have *instances* (the tokens themselves), and so only types give us a meaningful ratio of true to false beliefs.[8] A token process generates either a true or a false belief. If it made sense to speak of the reliability of token processes, then every token process would be either perfectly reliable or perfectly unreliable. Hence, the adequate bearers of reliability are not token processes. But any token process is an instance of indefinitely many types, whose reliability varies wildly. For instance, right now my belief that there is a computer in front of me is produced by a process belonging to all of these types: perceptually acquired belief; visually acquired belief; visually acquired belief in a subject with some degree of farsightedness; etc. Which of these types should be assessed for reliability in order to figure out whether the output belief is justified? The generality problem is just the generalization of this question to all cases of belief-formation.

Notice that although the generality problem questions the assumption behind the new evil demon problem and the blind reliability problem that we can unproblematically figure out the reliability of a belief-forming process, the formulation of the generality problem itself also makes assumptions as to how reliability is to be measured. In particular, usual formulations of the generality problem seem to assume that the reliability of a process is a matter of the ratio of true to false beliefs generated by that process. But, even assuming that the generality problem is solved, this conception of how to measure reliability is problematic. This is the fourth problem for Reliabilism that I will consider.

There is a lottery problem that arises for Reliabilism under the assumption that reliability is to be measured by the ratio of true to false beliefs produced by a process. For consider a large lottery, say one with 1,000,000 numbers in it, and suppose that you have one ticket. Are you justified in believing that your ticket will lose, based solely on the extremely low chances it has of winning? Many philosophers would say that you do not *know* that your ticket will lose, and some would even say that you are not even justified in believing it.[9] But surely you can become justified in believing that your ticket lost by some other means, say by reading about the winner in the newspaper. Now, depending on what the Reliabilist wants to say about believing that your ticket will lose based solely on the chances, he may have an obvious or a less obvious problem.

The obvious problem arises if the Reliabilist wants to say you are not justified in believing that your ticket will lose based solely on the chances. For consider the fact that newspapers (or websites) contain misprints way more often than once in a million times.[10] When you believe that your ticket lost by reading about it in the

[8] Comesaña (2006) argues against this assumption that we can make sense of the reliability of a token process, but also notes that this will not help Reliabilists avoid the generality problem.

[9] This is related to a different lottery problem, the one raised by Kyburg (1961) on the issue of how credences are related to full beliefs.

[10] Consider a daily newspaper with about 100,000 words per issue. For its rate of misprinted words to be one in a million, it would have to contain at most one misprint every ten days. Or consider a hundred newspapers that put out one lottery result every day. For the rate of error to be one in a million there would have to be no more than a single error in all the newspapers in twenty-eight years.

newspaper, you infer that your ticket did not win via your belief that ticket number *n* won and that your ticket is not *n*. This inferentially acquired belief will be justified, according to the Reliabilist, just in case the process is conditionally reliable (which it is in this case) and all of the input beliefs are themselves justified. So, in order to deliver the verdict that your belief that your ticket lost is in this case justified the Reliabilist is committed to saying that you can become justified in believing that ticket *n* won the lottery by reading about it in the newspaper. Let us assume that the process that produces that belief is a belief-independent process. This assumption seems false in this case (surely your background beliefs about the kind of newspaper you are reading plays some role here), but lifting it will not help the Reliabilist. Bracketing for the time being the generality problem, that belief-independent process is something like the following: believing that ticket number *n* won the lottery by reading it in this kind of newspaper. The reliability of that process, measured as a ratio of true to false beliefs, is surely lower than $\frac{999{,}999}{1{,}000{,}000}$. So, in assuming that you can come to know your ticket lost by reading about it in the newspaper, the Reliabilist is committed to thinking that a reliability ratio of less than $\frac{999{,}999}{1{,}000{,}000}$ is sufficient for a process to produce justified beliefs. This is in direct conflict with the claim that you cannot know that your ticket lost based solely on the chances.[11]

If the Reliabilist does say that you are justified in believing that your ticket will lose based solely on the chances, then that direct problem doesn't arise. But there is still a problem. For, although this is not part of the official Reliabilist story, it is very natural to assume that degrees of justification co-vary with degrees of reliability. If so, the Reliabilist will still have to face a problem regarding lotteries. For the reliability of the process of believing that your ticket lost based on reading in the newspaper that some other ticket won is still much lower than the reliability of the process of believing that your ticket lost based solely on the chances, and yet surely the first belief is at least as well as justified as (and, actually, better justified than) the second one.[12]

To recap, Reliabilism faces two different kinds of problems. One set of problems has to do with the material adequacy of the theory. The new evil demon problem challenges the necessity of reliability for justification, whereas the blind reliability problem challenges its sufficiency. The second set of problems has to do with the concept of reliability itself. The generality problem points out that the

[11] It might seem relevant that, even if the newspaper misprinted the winner, the chances that your number was the real winner are still very low. But the important point is that they are not lower than one in a million. Now, perhaps some misprints are more likely than others, and in particular numbers close to the misprinted one are more likely to be the winner than numbers further apart from it (where the distance is measured, say, in number of incorrect digits). Still, the leftover chance of an egregious misprint seems still higher than one in a million, and if it isn't we can just make the lottery larger.

[12] Recall that this is a problem if we are measuring reliability in terms of truth-ratios. In Comesaña (2009) I argue that understanding reliability in terms of conditional probability solves this particular problem.

Reliabilist needs to give us a principled way of selecting a unique type from among the indefinitely many ones that any token process belongs to, whereas the lottery problem means that a natural way of measuring reliability (via truth-falsehood ratios) will not do.

A different kind of objection is that whereas Reliabilism as developed can account (perhaps) for the justification of beliefs, it is not clear how to adapt it to other doxastic attitudes. Within the realm of coarse-grained epistemology, we could try the following: just as a belief is justified if the process that produced it is reliable, so too disbelief is justified if the process is anti-reliable, and suspension of judgment is justified if the process is neither reliable nor anti-reliable. And within fine-grained epistemology, we could perhaps try the following: a degree of belief (or credence) is justified if and only if it matches the degree of reliability of the process that produced it. Those options might well work,[13] but they raise the same issue we touched upon above: how to measure the reliability of a belief-forming process-type. Before tackling that issue, however, we need a solution to the generality problem.

9.4 How to Solve the Generality Problem

Alston (1988) proposes a version of Reliabilism according to which a belief is justified just in case it is based on an adequate ground, where the adequacy of a ground is a matter of its reliability. In place of Alston's grounds, we can invoke the notion of evidence. This will give us a principled way of selecting a type for each belief-forming process-type: the type *believing that p on the basis of e*. Remember that Evidentialists themselves appeal to this notion of basing in their account of well-foundedness. It is *that type* that should be assessed for reliability. This is what I proposed in Comesaña (2006). My proposal was the following:

Well-Founded Reliabilism (first pass): A belief that *p* by *S* is epistemically justified if and only if:

(i) *S* has evidence *E*;
(ii) the belief that *p* by *S* is based on *E*; and
(iii) the type *producing a belief that p based on evidence E* is a reliable type.

That proposal, however, ignores the reasoning behind clause (ii)(c) of the Evidentialist definition of well-foundedness. Recall the reason: if I base my belief on a subset of my evidence that justifies it, but ignoring some other evidence that

[13] Although see Tang (2016a) for more on how Reliabilists should capture suspension of judgment.

I have which does not justify it, then my belief is not well-founded. Taking that into account yields the following refined version of Well-founded Reliabilism:

Well-founded Reliabilism: A belief that *p* by *S* is epistemically justified if and only if:

1. *S* has evidence *E*;
2. the belief that *p* by *S* is based on *E*;
3. the type *producing a belief that p based on evidence E* is a reliable type; and
4. there is no more inclusive body of evidence *E'* had by *S* at *t* such that the type *producing a belief that p based on evidence E'* is not a reliable type.

Notice that if we bracket clause 2, what is left is a definition of propositional justification for Reliabilism.

When assessing Evidentialism, we said that the theory needed an account of evidence and its possession. Well-founded Reliabilism inherits those needs, and the same options canvassed earlier are available here as well. In fact, however, a further development of Well-founded Reliabilism will provide us with an argument for a specific account of evidence. I turn to that development next.

9.5 How to Measure Reliability

We have a solution to the generality problem insofar as we have a principled way of selecting a process type for any token process of belief formation. But this doesn't yet give us a complete Reliabilist theory, for we need to figure out how to measure the reliability of a process type.

One possible answer, suggested above, is to think of the reliability of a process as the truth-to-falsity ratio of its outputs. There are two varieties of this truth-ratio conception of reliability: reliability as actual high truth-ratio, or reliability as counterfactual high truth-ratio. The first one counts a process as reliable if and only if the ratio of truths to falsehoods within its actual outputs is sufficiently high, whereas the second one concerns not just actual truth-ratios but counterfactual ones as well. A counterfactual conception is preferable insofar as there may be processes that, although intuitively justification-conferring, do not have many actual outputs, and so its actual truth-ratio may be coincidentally low. Shifting to a counterfactual account of reliability is not without issues, however, because one needs to restrict the relevant range of counterfactual applications in order to not trivialize the account. This issue with the counterfactual account is reminiscent of the generality problem, for we need to decide which of the counterfactual applications of the process are the relevant ones to measure reliability. Moreover, any truth-ratio account of how to measure reliability runs into the lottery problem canvassed above.

A different measure of reliability has also been proposed by Alston (and used in Comesaña (2009)): reliability as high conditional probability. The idea here is that a type of the form *believing that p based on e* is reliable if and only if, for some suitable probability function *Pr* and some suitable threshold *r*, $Pr(p|e) \geq r$. Two immediate questions about that approach are: what is a suitable probability function, and what is a suitable threshold? A third issue has to do with Carnap's distinction between confirmation as firmness and confirmation as increase in firmness.[14] $Pr(p|e)$ can be greater than *r* even if *e* doesn't raise the probability of *p*—indeed, for $r < 1$ it can happen that $r < Pr(p|e) < Pr(p)$. If this is the case, then we can hardly say that it is *e* that justifies the subject's belief that *p*. This need not be a problem for the Reliabilist who appeals to the notion of evidence only to solve the generality problem, and who doesn't take himself to be elucidating the notion of evidential justification, but only justification simpliciter. It may, however, be an issue for the Evidentialist—we'll come back to this point later.

Let us leave those questions aside for the moment, and formulate the resulting version of Reliabilism:

Probabilistic Evidentialist Reliabilism: A belief that *p* by *S* is justified if and only if:

1. *S* has evidence *E*;
2. the belief that *p* by *S* is based on *E*;
3. $Pr(p|E) \geq r$.
4. There is no more inclusive body of evidence *E*' had by *S* such that $Pr(p|E') < r$.

The resulting theory is called "Probabilistic Evidentialist Reliabilism" to honor the three components that play a crucial role in it: evidence, reliability, and probability. The obvious question at this point, however, is what notion of probability is being used here. I turn now to that question.

9.6 What is Pr?

The theory we have arrived at is very similar to one recently proposed by Tang (2016b), which in turn, as noticed by Pettigrew (forthcoming), is at least extensionally equivalent to a theory proposed by Dunn (2015) (at least when Dunn's theory is supplemented as we have done here to deal with the generality problem) as well as to Pettigrew's own theory. Those theories do have an answer to the question about the notion of probability being used. All three of Tang, Pettigrew, and Dunn answer that it is an objective, non-evidential probability that is in question. It is objective because it does not measure the actual degrees of belief of

[14] See the preface to the second edition of Carnap (1950).

any particular agent. But it is not an evidential probability function, because it is not a measure of the degrees of belief it is rational to have. Or, more precisely, it will turn out to be an evidential probability function in the end (after all, the theories are all theories of justified credence), but only because the theories have it that the evidential probability function just coincides with a non-evidential one. The theories are trying to explain evidential probabilities, and they do so by appealing to a non-evidential type of probability. In doing this, they honor Goldman's constraint, according to which an adequate theory of epistemic justification must explain such justification in non-epistemic terms.

Pettigrew is the most explicit about this. He says:

> Given Jenann Ismael's distinction between single-case objective probabilities and general objective probabilities, our notion falls under the latter heading (Ismael (2011)). They are what Elliott Sober (2010) calls macro-probabilities; they are what David Albert (2000) finds in classical statistical mechanics. For Ismael, single-case probabilities tend to be unconditional, and assign values to particular token events, such as this particular die landing six on this particular roll. They are sometimes called chances, and they are the sort of probabilities we find in quantum mechanics. They are the objective probabilities that propensity accounts and best-system analyses aim to explicate. General probabilities, in contrast, tend to be conditional, and they take as arguments a pair of event types, such as dice of a certain sort landing six given that they are rolled. These are the sorts of probabilities that are found in statistical mechanics and evolutionary biology. They are what frequentist accounts attempts to explicate.

And he goes on to provide some examples:

> Crucially, for our purpose, non-trivial general probabilities are possible even in deterministic worlds, whereas non-trivial chances are not—indeed, it is part of what it means for a world to be deterministic that the chance of an event at any time is either 0 or 1. Thus, in such a world, any particular roll of any particular die either is determined to come up six or is determined not to come up six. Nonetheless, it is still possible in such a world that the general objective probability of a die with certain general physical properties landing six given that it is rolled is $\frac{1}{6}$. Similarly, while it is determined by the deterministic laws whether any particular egg in any human reproductive system will survive to reproductive age or will not, there is nonetheless a non-trivial probability that an egg will survive to reproductive age, given that it is a human egg—and this is a general probability. And while it is determined by the deterministic laws whether any particular block of ice in warm water will or will not melt, there is nonetheless a non-trivial (though very high) probability that a block of ice will melt given that it is in warm water—again, this is a general probability.

Given that this is the nature of the probability function *Pr*, however, it is quite clear that it will render all of the theories that appeal to it materially inadequate—they will all have clear counterexamples. Moreover, the counterexamples will not be easily brushed off as marginal or somehow not terribly relevant—they strike at the heart of the theories, and show that they are, simply put, wrong.

BonJour's clairvoyant counterexample to Reliabilism was designed to show that reliability is not sufficient for justification. I argued (in Comesaña (2010a)) that a kind of Reliabilism that incorporates Evidentialist themes is not vulnerable to BonJour's counterexamples. However, counterexamples of the same kind can be given even against sophisticated Reliabilist theories that incorporate Evidentialist notions, provided that they appeal to the kind of objective probability that Pettigrew is alluding to. BonJour's counterexamples rested on the fact that blind reliability doesn't provide justification. Adding the notion of evidence to Reliabilism, I thought, provided a way out of the blindness. But measuring reliability in terms of general objective probabilities blinds Reliabilism once again, even when inoculated with evidence.

Suppose that you know by seeing that a certain leaf has shape *S*. As it happens, the general objective probability of its being the leaf of an oak tree given that it has shape *S* is *n*. You, however, are no botanist, and know nothing about oaks and their leaves. Even if, by chance, you happen to assign credence *n* to the proposition that the leaf is an oak leaf, that credence will in no way be justified. Or suppose that you know that a certain patient has symptoms *S*. As it happens, the general objective probability of a person's having disease *D* given that they exhibit symptoms *S* is *m*. You, however, are no physician, and know nothing about disease *D* and its symptoms. Even if, by chance, you happen to assign credence *m* to the proposition that the patient exhibiting symptoms *S* has disease *D*, that credence will in no way be justified. Moreover, those credences will not be justified even if your assignment of credences is done on the basis of the evidence in question. Suppose that you base your credence about the leaf coming from an oak on the fact that it has shape *S*, and you base your credence in the patient's having disease *D* on the fact that he exhibits symptoms *S*. Far from this making your credence assignments rational, it just highlights the role that blind luck is playing in your credence assignments—a kind of luck incompatible with justification.

Now, it may be replied that if you have no idea about the connection between the leaf shape and its provenance, or between the symptoms and its causes, then you are not really basing your credence assignments on the evidence. At best, the evidence is playing a merely causal role in those credence assignments—but even advocates of causal accounts of basing need to grant that the existence of a causal relation is not *sufficient* for epistemic basing. Perhaps the subject also needs to somehow appreciate the connection between the evidence and the doxastic attitude it justifies.

But this response cannot help the Evidentialist Reliabilist who appeals to an objective type of probability. To begin with, it has a decidedly internalist flavor which no Reliabilist worth the name (not even an Evidentialist one) should be comfortable with. But, more seriously, if we buy the idea that basing requires an appreciation of the bearing of the evidence on the target doxastic attitude, not any old appreciation will do—it will have to be a *justified* appreciation. For instance, if you also just happen to think that n percent of leaves with shape S belong to oaks, that will in no way make your credence assignment more justified. And if your appreciation of the bearing of the evidence has to itself be justified, then this launches a regress that ends either with some evidence giving you doxastic justification even absent appreciation of its bearing, or with some such appreciation being justified non-evidentially. But if the regress is resolved the first way, that just leaves the theory open to the original objection. For now we will have a case where the subject assigns a credence n to a proposition p on the basis of some evidence E, and while it is true that $Pr(p|E) = n$, interpreting Pr as a general objective probability, the subject has no idea about the connection between E and p. On the other hand, if the regress is resolved the second way, by saying that justified appreciations of the bearing of evidence on doxastic attitudes may themselves be non-evidentially based, that is just incompatible with the theories, which have the consequence that all justified attitudes are evidence-based.

Cohen's new evil demon objection to Reliabilism was designed to show that reliability is not necessary for justification. I argued (in Comesaña (2002)) for a solution to that objection based on a two-dimensional semantics for "reliable."[15] But the theories of Dunn, Tang, and Pettigrew do not incorporate that detail of my position. As such, they are vulnerable to Cohen's objection—or a similar one at least. For suppose that a subject lives in a "counter-inductive" environment. For instance, suppose that the environment puts evolutionary pressure on individual birds of the same species to have different colors.[16] In that case, the general objective probability of the hypothesis that all ravens are black diminishes as the number of observed black ravens grows, for if they were all black it wouldn't be likely that many of them would exist. Nevertheless, it would be irrational for a subject to become more and more convinced that not all ravens are black the more black ravens she observes.

Behind these specific counterexamples lies a more fundamental problem with appealing to this kind of objective probability, and that is that it is a contingent kind of probability. The value of a particular conditional probability of this kind depends on the contingent regularities that obtain in the world. As such, it will only be rational to match one's credences to those probabilities

[15] Cf. Ball and Bloome-Tillman (2013).

[16] Compare Titelbaum (forthcoming) on the "Hall of different-colored birds."

after one learns about the correlations. But this kind of learning is itself an epistemic achievement, and it arguably involves rational belief. Therefore, the rationality of the doxastic attitudes cannot be explained in terms of that kind of probability.

9.7 Evidential Probability

If *Pr* cannot be the kind of objective probability function that Dunn, Tang, and Pettigrew take it to be, then what is it? Well, how about the ur-prior, that is to say, the evidential probability function? As we know by now, the evidential probability function determines two things: what credence it is rational to assign to different hypotheses in the absence of any evidence for or against them, and (via the conditional evidential probability function) what credences it is rational to assign to propositions given certain evidence. It is also an objective kind of probability, but not contingent.

The theory we have arrived at, then, is the following. First, the evidence a subject has at *t* consists of those propositions the subject is basically justified in believing. Second, the credences a subject is justified in assigning at *t* are those determined by $Cr_u(-|E)$, where Cr_u is the ur-prior and *E* is the subject's evidence at *t*.

Let us now see how the appeal to the ur-prior can answer the problems for Reliabilism that we argued cannot be answered by appeal to an objective probability function. BonJour's style of counterexample focused on cases where reliability is allegedly not sufficient for justification. But if reliability is measured in terms of the ur-prior, then this kind of counterexample is impossible. For, simply put, if a subject is not justified in believing a proposition *p* despite having evidence that makes *p* sufficiently likely according to a probability function *Pr*, then *Pr* is not the ur-prior. Similarly, Cohen's counterexamples require a situation where a subject is justified in believing a proposition *p* even though the subject's evidence makes *p* unlikely according to *Pr*—but, again, that is only possible if *Pr* is not the evidential probability function. More generally, as I argued in section 9.6, both BonJour's and Cohen's counterexamples to Reliabilism rely on measuring reliability according to a function whose values are contingent. But the values delivered by the ur-prior are not contingent.[17]

[17] In Comesaña (2010a) I argued that the mere appeal to evidence could answer BonJour's counterexample. To be more precise, I granted that maybe BonJour's counterexamples did show that reliability is only necessary for justification, but I didn't comment on the fact that this just means that Reliabilism thus conceived was at best only a partial account of evidential fit. In that same paper I adopted my previous answer to Cohen's new evil demon problem presented in Comesaña (2002). In effect, my proposal there is one way to make contingent reliable connections into necessary ones. Given the necessity of evidential probabilities, this more roundabout solution is not necessary.

Now, in one respect, the fact that measuring reliability according to the ur-prior relieves Reliabilism of its problems with counterexamples is, of course, a welcome feature. In another respect, however, it might seem that the appeal to the evidential probability function makes the Reliabilist answer to alleged counterexamples *too easy*. Relatedly, there is a circularity worry here: we are trying to give an account of epistemic justification, and we end up appealing to the ur-prior, which just encodes under what conditions doxastic attitudes are justified. Moreover, we are not giving an independent specification of the ur-prior, but just appealing to it, whatever it is. How, therefore, is this progress? I come back to this question below. First, however, we need to make explicit the formulation of the theory with respect to credences, and not just beliefs.

9.8 Credences

The theory we have arrived at is the following: experiences provide the subject with an initial corpus of propositions as evidence, provided that they justify the subject in believing their contents; this initial corpus of evidence can then justify belief in a further proposition when the conditional probability of that proposition given a suitable subset of the initial corpus is high enough. A natural thought here would be to think that propositions justified downstream of experience can then join forces with the basically justified propositions and justify still more propositions, the edifice of justified propositions growing under its own steam, so to speak. But although this picture is a very traditional one, my Mere Lemmas principle from chapter 8 entails that it cannot be the literal truth.[18]

The resulting theory can be formulated as follows:[19]

Coarse-Grained Evidentialist Reliabilism: A belief that p by S is justified if and only if:
Either:

1. S's experiences provide him with p; or
2a. S's experiences provide him with E;
2b. the belief that p by S is based on E;
2c. $Cr_u(p|E) \geq r$;
2d. There is no more inclusive body of evidence E' had by S such that $Pr(p|E+) < r$.

[18] Sosa (2016) criticized Evidentialist Reliabilism precisely on the basis that it, together with Evidentialism, assumed that all beliefs are evidentially justified. As the theory of evidence developed in this book shows, I agree with Sosa.

[19] The definitions are implicitly relativized to a time. That doesn't mean that the theory is a version of "time-slice" epistemology, according to which which doxastic attitudes are justified at a time supervenes on the subject's mental states at a time, for it leaves it open that past experiences may provide subjects with present evidence.

I call the resulting theory "Coarse-Grained Evidentialist Reliabilism" because it accounts only for the rationality of coarse-grained doxastic attitudes, and not of fine-grained ones. It also incorporates a naïve Lockean conception of the connection between credences and beliefs. One may want to replace that with a contextualist or subject-sensitive invariantist version of Lockeanism, or forego Lockeanism altogether. Notice also that I appeal to the notion of an experience providing a proposition as a reason. As explained in chapter 6, this will happen whenever the justification that the subject has for believing the content of the experience is not defeated by other beliefs the subject is justified in having.

My main interest in this book, however, has not been with the epistemology of full belief, but with the epistemology of credences. In that respect, Coarse-Grained Evidentialist Reliabilism is inferior to plain old Evidentialism, which applied to all doxastic attitudes. There is a very natural way, however, to transform Coarse-Grained Evidentialist Reliabilism into a theory that applies to credences:

Fine-Grained Evidentialist Reliabilism: A credence x in p by S is justified if and only if:
Either:

1. $x = 1$ and p is either a logical truth or S's experiences provide him with p; or
2a. S's experiences provide him with E;
2b. S's credence x in p is based on E;
2c. $Pr(p|E) = x$.
2d. There is no more inclusive body of evidence E' had by S such that $Pr(p|E') \neq x$.

What of the complaint, raised before, that the theory conflates justification as firmness with justification as increase in firmness? The complaint doesn't apply to Fine-Grained Evidentialist Reliabilism because it does not deal with a threshold notion of justification.

9.9 Conclusion

Fine-Grained Evidentialist Reliabilism just is the theory developed in this book. That theory can be approached from two directions. In the previous chapters in this book I developed it as arising from judgments about practical rationality and Fumerton's thesis, and in contrast to both Psychologism and Factualism. In this chapter, I showed how the very same theory can be developed as arising from the insights of both Evidentialism and Reliabilism. But is Fine-Grained Evidentialist Reliabilism well-named? It is in the sense of honoring its ancestry, but the resulting view is, in important respects, neither Evidentialist nor Reliabilist. (The name is also, of course, more than a mouthful.)

Starting with Evidentialism, the most fundamental difference between the views is that Fine-Grained Evidentialist Reliabilism embraces the possibility of non-evidentially justified beliefs. Relatedly, Fine-Grained Evidentialist Reliabilism has it that a subject's evidence is constituted exclusively by those propositions provided as evidence by his experiences (remember that we are treating "experiences" as somewhat of a placeholder for all non-factive mental states that can provide evidence). For the Evidentialist and for the proponent of Psychologism, remember, there are two fundamentally different kinds of evidence: evidence can consist of propositions, which are had as evidence only if justifiedly believed, or of experiences, for which there is no distinction between evidence and its possession. Fine-Grained Evidentialist Reliabilism, on the other hand, has a unified conception of evidence and its possession. Conversely, whereas for the Evidentialist all propositions are evidentially justified, for the Fine-Grained Evidentialist Reliabilist some propositions are justified but not on the basis of any evidence. Both kinds of theory therefore posit some kind of bifurcation: one in the notion of evidence and its possession, the other on the ways beliefs can be justified.

Fine-Grained Evidentialist Reliabilism replaces the Evidentialist notion of fit with an appeal to the ur-prior. So, in this respect, while Fine-Grained Evidentialist Reliabilism might not be better off than Evidentialism, it clearly isn't worse off. I conceived of the Reliabilist part of Evidentialist Reliabilism as providing an answer to the question of fit, but I didn't pay sufficient attention to the question of how to measure reliability. If we measure reliability by an objective probability function, then the old counterexamples to Reliabilism come back with a vengeance, and the admixture of Evidentialism will not help. Therefore, while the versions of Reliabilism advocated by Dunn, Tang, and Pettigrew do indeed provide us with a non-circular account of justification, that account is simply materially inadequate. Moreover, its material inadequacy can be traced back precisely to the fact that they conceive of *Pr* as a contingent function, and as such the only justified way to match our credences to it is by learning about those contingent correlations. But part of what we need to explain when we explain epistemic justification is precisely how it is that we are justified in learning about those correlations. An explicit part of Goldman's project in epistemology is to provide a *reductivist* account of epistemic justification. Insofar as the kind of Evidentialist Reliabilism defended here appeals to the notion of evidential probability, it cannot fulfill that promise. As I see it, however, reductivism of that kind comes at the cost of material inadequacy. The resulting theory, therefore, is no more Reliabilist than, say, Carnap's theory was.

10
Conclusion

The received view about the relationship between experience and justified belief was, until recently, the view that experience functions as evidence for belief (I called this view "Psychologism"). A more recent picture (receiving a classic expression in Williamson's *Knowledge and Its Limits*, but anticipated by disjunctivists such as Hinton (1967), Snowdown (1980), and McDowell (1982)) is that experience provides, but does not consist in, evidence. According to this view, moreover, experience provides evidence only when (and because) it provides knowledge (I called this view "Factualism"). In this book I have developed and defended a third kind of theory, Experientialism. Experientialism agrees with Factualism in claiming that experience provides, but does not consist in, evidence. However, Experientialism differs from Factualism in holding that experience provides evidence not only when it provides knowledge, but when it provides (basically) justified belief as well.

My theory would end up agreeing with Factualism anyway if justified belief were just identical with knowledge. But, as I have also argued in this book, justification is not identical with knowledge. A crucial argument for both theses (for Experientialism and for the claim that justified belief does not coincide with knowledge) concerns connections between practical and theoretical rationality. It is sometimes rational to act on the basis of false beliefs, but it is never rational to act on the basis of unjustified beliefs. It follows that there are justified false beliefs. Moreover, the beliefs on the basis of which it is rational to act are those that are provided as evidence by experience. Evidence, in my theory, is constituted by those propositions that determine which possible states we should consider when deciding what to do: we should consider exactly those states compatible with our evidence. Seen in this light, both Psychologism and Factualism distort practical as well as theoretical reasoning. According to Psychologism, only our experiences are evidence, and according to Factualism only what we know is evidence.[1] Thus, when Tomás has an experience as of Lucas offering some candy to him (when in fact Lucas is only offering a marble), both Psychologism and Factualism would have it that Tomás should consider the possibility that Lucas is offering a marble.

[1] It is basic evidence that is relevant here. Psychologists could claim that we have more evidence than our experiences, but we cannot treat that further evidence as playing the role of determining which states to consider when deciding what to do without running afoul of the Mere Lemmas principle discussed in chapter 8.

Being Rational and Being Right. Juan Comesaña, Oxford University Press (2020). © Juan Comesaña.
DOI: 10.1093/oso/9780198847717.001.0001

But not only does Tomás not consider this possibility, he is right not to consider it (at least until he accumulates some evidence for it).

Because of my reliance on the connections between practical and theoretical rationality I framed Experientialism within the confines of traditional decision theory, which I explained in chapter 2. As I also mentioned in that chapter, I do not take any form of Bayesian decision theory to be the literal truth, for we can be justified in having probabilistically incoherent credences. However, Bayesian decision theory is a useful idealization in that it allows us to concentrate on the peculiar contribution that our ignorance of matters of facts makes to the rationality of our actions. Within the confines of that idealization, I argued in chapter 3 against a Subjective and for an Objective variety of Bayesianism. One of the most influential versions of Objective Bayesianism incorporates Williamson's equation of our evidence with what we know, thus turning Objective Bayesianism into a form of Factualism. In chapters 5 through 7 I argued against Factualism (and Psychologism) and for Experientialism, along the lines described at the beginning of this chapter. Chapters 7 through 9 took up some loose ends: what to say about cases where we already have unjustified attitudes (chapter 7), the problem of easy rationality (chapter 8), and the relationship between Experientialism and some traditional epistemological theories, such as Evidentialism, Reliabilism, and my own Evidentialist Reliabilism (chapter 9).

I am not the first to defend a view along the lines of Experientialism: Dancy (2000), Schroeder (2008), Schroeder (forthcoming), and Fantl and McGrath (2009) are among my predecessors and fellow travelers. My contribution in this book has been to develop such a theory in detail, embed it in the framework of traditional decision theory, argue for it by way of the connections between practical and theoretical rationality, and argue for its superiority to both Psychologism and Factualism. I hope to have convinced you that the view is worth taking seriously.

References

al-Hibri, Azizah. 1978. *Deontic Logic: A Comprehensive Appraisal and a New Proposal.* University Press of America.

Albert, David. 2000. *Time and Chance.* Harvard University Press.

Alston, William. 1986. "Epistemic Circularity." *Philosophy and Phenomenological Research* 47: 1–30.

Alston, William. 1988. "An Internalist Externalism." *Synthese* 74: 265–83.

Armendt, Brad. 1993. "Dutch Books, Additivity, and Utility Theory." *Philosophical Topics* 21 (1).

Armstrong, David. 1978. *Universals and Scientific Realism.* Cambridge University Press.

Arntzenius, Frank. 2003. "Some Problems for Conditionalization and Reflection." *Journal of Philosophy* 100 (7): 356–70.

Bacon, Andrew. 2014. "Giving Your Knowledge Half a Chance." *Philosophical Studies* 171: 373–97.

Ball, Brian and Michael Bloome-Tillmann. 2013. "Indexical Reliabilism and the New Evil Demon." *Erkenntnis* 78: 1317–36.

Bergmann, Michael. 2006. *Justification without Awareness.* Oxford University Press.

Bird, Alexander. 2007. "Justified Judging." *Philosophy and Phenomenological Research* 74 (1): 81–110.

BonJour, Lawrence. 1980. "Externalist Theories of Empirical Knowledge." *Midwest Studies in Philosophy* 5: 53–73.

Bronfman, Aaron and Janice Dowell. 2018. "The Language of Reasons and 'Ought'." In *Oxford Handbook of Reasons and Normativity*, edited by Daniel Star, 85–112. Oxford University Press.

Broome, John. 1999. "Normative Requirements." *Ratio* 12: 398–419.

Broome, John. 2013. *Rationality whrough Reasoning.* Wiley-Blackwell.

Byrne, Alex. 2016. "The Epistemic Significance of Experience." *Philosophical Studies* 173: 947–67.

Calabrese, Philip. 1987. "An Algebraic Synthesis of the Foundations of Logic and Probability." *Information Sciences*, no. 42: 187–237.

Cantwell, John. 2008. "Changing the Modal Context." *Theoria* 74: 331–51.

Carnap, Rudolph. 1950. *Logical Foundations of Probability.* University of Chicago Press.

Carnap, Rudolph. 1971. "A Basic System of Inductive Logic I." In *Studies in Inductive Logic and Probability*, edited by Richard Jeffrey, 1: 33–165. University of California Press.

Carnap, Rudolph. 1980. "A Basic System of Inductive Logic Ii." In *Studies in Inductive Logic and Probability*, edited by Richard Jeffrey, 2: 7–155. University of California Press.

Casullo, Albert. 2018. "Pollock and Sturgeon on Defeaters." *Synthese* 195 (7): 2897–906.

Chellas, Brian. 1974. "Conditional Obligation." In *Logical Theory and Semantic Analysis*, edited by Soeren Stenlund, 23–33. Dordrecht.

Chisholm, Roderick. 1963. "Contrary-to-Duty Imperatives and Deontic Logic." *Analysis* 24 (2): 33–6.

Chisholm, Roderick. 1966. *Theory of Knowledge.* 1st edition 1966; 2nd edition 1977; 3rd edition 1989. Prentice Hall.

Christensen, David. 1996. "Dutch-Book Arguments De-Pragmatized: Epistemic Consistency for Partial Believers." *Journal of Philosophy* 93 (9): 450–79.
Christensen, David. 2004. *Putting Logic in Its Place*. Oxford University Press.
Christensen, David. 2007a. "Does Murphy's Law Apply in Epistemology? Self-Doubt and Rational Ideals." In *Oxford Studies in Epistemology*, edited by Tamar Gendler and John Hawthorne, 2: 3–31. Oxford University Press.
Christensen, David. 2007b. "Epistemology of Disagreement: The Good News." *Philosophical Review* 116 (2): 187–217.
Cohen, Stewart. 1984. "Justification and Truth." *Philosophical Studies* 46: 279–95.
Cohen, Stewart. 1988. "How to Be a Fallibilist." *Philosophical Perspectives* 2: 91–123.
Cohen, Stewart. 2002. "Basic Knowledge and the Problem of Easy Knowledge." *Philosophy and Phenomenological Research* 65 (2): 309–29.
Cohen, Stewart. 2010. "Bootsrapping, Defeasible Reasoning and *a Priori* Justification." *Philosophical Perspectives* 24 (1): 141–59.
Cohen, Stewart, and Juan Comesaña. 2013a. "Williamson on Gettier Cases and Epistemic Logic." *Inquiry* 56 (1): 15–29.
Cohen, Stewart, and Juan Comesaña. 2013b. "Williamson on Gettier Cases in Epistemic Logic and the Knowledge Norm for Rational Belief: A Reply to a Reply to a Reply." *Inquiry* 56 (4): 400–15.
Cohen, Stewart, and Juan Comesaña. forthcoming. "Rationality and Truth." In *The New Evil Demon*. Oxford University Press.
Comesaña, Juan. 2002. "The Diagonal and the Demon." *Philosophical Studies* 110: 249–66.
Comesaña, Juan. 2005a. "Justified vs. Warranted Perceptual Belief: Resisting Disjunctivism." *Philosophy and Phenomenological Research* 71 (2): 367–83.
Comesaña, Juan. 2005b. "Unsafe Knowledge." *Synthese* 146: 395–404.
Comesaña, Juan. 2005c. "We Are (Almost) All Externalists Now." *Philosophical Perspectives* 19 (1): 59–76.
Comesaña, Juan. 2006. "A Well-Founded Solution to the Generality Problem." *Philosophical Studies* 129: 127–47.
Comesaña, Juan. 2009. "What Lottery Problem for Reliabilism?" *Pacific Philosophical Quarterly* 90 (1): 1–20.
Comesaña, Juan. 2010a. "Evidentialist Reliabilism." *Noûs* 44 (4): 571–600.
Comesaña, Juan. 2010b. "Reliabilism." In *The Routledge Companion to Epistemology*, edited by Sven Bernecker and Duncan Pritchard, 176–86. Routledge.
Comesaña, Juan. 2013a. "Reply to Pryor." In *Contemporary Debates in Epistemology*, edited by Matthias Steup Ernest Sosa and John Turri, 239–43. Blackwell.
Comesaña, Juan. 2013b. "There Is No Immediate Justification." In *Contemporary Debates in Epistemology*, edited by Matthias Steup Ernesto Sosa and John Turri: 222–34. Blackwell.
Comesaña, Juan. 2016. "Normative Requirements and Contrary-to-Duty Obligations." *Journal of Philosophy* 112 (11): 600–26.
Comesaña, Juan. 2017. "On Sharon and Spectre's Argument against Closure." *Philosophical Studies* 174 (4): 1039–46.
Comesaña, Juan. 2018a. "A Note on Knowledge-First Decision Theory and Practical Adequacy." In *Pragmatic Encroachment in Epistemology*, edited by Matthew McGrath and Brian Kim, 206–11. Routledge.
Comesaña, Juan. 2018b. "Whither Evidentialist Reliabilism?" In *Believing in Accordance with the Evidence*, edited by Kevin McCain, 307–25 Springer.
Comesaña, Juan. forthcoming. "Empirical Justification and Defeasibility." *Synthese*.

Comesaña, Juan, and Matthew McGrath. 2014. "Having False Reasons." In *Epistemic Norms*, edited by Clayton Littlejohn and John Turri, 59–80. Oxford University Press.
Comesaña, Juan, and Matthew McGrath. 2016. "Perceptual Reasons." *Philosophical Studies* 173 (4): 991–1006.
Comesaña, Juan, and Carolina Sartorio. 2014. "Difference-Making in Epistemology." *Noûs* 48 (2): 368–87.
Comesaña, Juan, and Eyal Tal. 2015. "Evidence of Evidence Is Evidence (Trivially)." *Analysis* 75 (4): 557–9.
Conee, Earl, and Richard Feldman. 1985. "Evidentialism." *Philosophical Studies* 48 (1): 15–34.
Conee, Earl, and Richard Feldman. 1998. "The Generality Problem for Reliabilism." *Philosophical Studies* 89: 1–29.
Conee, Earl, and Richard Feldman. 2001. "Internalism Defended." In *Epistemology: Internalism and Externalism*, edited by Hilary Kornblith, 231–60. Blackwell.
Dancy, Jonathan. 2000. *Practical Reality*. Oxford University Press.
Danielson, Sven. 1968. *Preference and Obligation: Studies in the Logic of Ethics*. Filosofiska föreningen.
DeCew, Judith Wagner. 1981. "Conditional Obligation and Counterfactuals." *Journal of Philosophical Logic* 10: 55072.
Dowell, Janice. 2012. "Contextualists Solutions to Three Puzzles about Practical Conditionals." In *Oxford Studies in Metaethics*, edited by Russ Shaffer-Landau, 7: 271–303. Oxford University Press.
Dretske, Fred. 1970. "Epistemic Operators." *Journal of Philosophy* 67: 1007–23.
Duke, Annie. 2018. *Thinking in Bets*. Portfolio/Penguin.
Dummett, Michael. 1959. "Truth." *Proceedings of the Aristotelian Society* 59. Reprinted in his *Truth and Other Enigmas*, Duckworth, 1978.
Dunn, Jeff. 2015. "Reliability for Degrees of Belief." *Philosophical Studies* 172 (7): 1929–52.
Dutant, Julien. forthcoming. "Knowledge-Based Decision Theory and the New Evil Demon." In *The New Evil Demon*, edited by Julien Dutant and Fabian Dorsch. Oxford University Press.
Egan, Andy. 2007. "Some Counterexamples to Causal Decision Theory." *Philosophical Review* 116 (1): 93–114.
Elga, Adam. 2010. "Subjective Probabilities Should Be Sharp." *Philosopher's Imprint* 10 (5): 1–11.
Fantl, Jeremy, and Matthew McGrath. 2002. "Evidence, Pragmatics, and Justification." *Philosophical Review* 112 (1): 47–67.
Fantl, Jeremy, and Matthew McGrath. 2007. "On Pragmatic Encroachment in Epistemology." *Philosophy and Phenomenological Research* 75 (3): 558–89.
Fantl, Jeremy, and Matthew McGrath. 2009. *Knowledge in an Uncertain World*. Oxford University Press.
Feldman, Richard. 2003. *Epistemology*. Prentice Hall.
Feldman, Richard. 2007. "Reasonable Religious Disagreements." In *Philosophers without God: Medtiations on Atheism and the Secular Life*, edited by Louise Antony, 194–214. Oxford University Press.
Fine, Kit. 1975. "Review of Counterfactuals, by David Lewis." *Mind* 84: 341–4.
Fitch, Frederick B. 1963. "A Logical Analysis of Some Value Concepts." *The Journal of Symbolic Logic* 28 (2): 135–42.
Fitelson, Branden. 1999. "The Plurality of Bayesian Measures of Confirmation and the Problem of Measure Sensitivity." *Philosophy of Science* 66: S362–78.
Fitelson, Branden. 2005. "Inductive Logic." In *Philosophy of Science: An Encyclopedia*, edited by J. Pfeifer and S. Sarkar, 384–94. Routledge.

Fitelson, Branden. 2008. "A Decision Procedure for Probability Calculus with Applications." *The Review of Symbolic Logic* 1: 111–25.

Forrester, James. 1984. "Gentle Murder, or the Advervial Samaritan." *Journal of Philosophy* 81: 193–6.

Fumerton, Richard. 1990. *Reason and Morality*. Cornell University Press.

Fumerton, Richard. 1996. *Metaepistemology and Skepticism*. Rowman; Littlefield.

Gallow, J. D. 2013. "How to Learn from Theory-Dependent Evidence; or Commutativity and Holism: A Solution for Conditionalizers." *British Journal for Philosophy of Science* 65: 493–529.

Garber, Daniel. 1983. "Old Evidence and Logical Omniscience in Bayesian Confirmation Theory." In *Minnesota Studies in the Philosophy of Science*, edited by John Earman, 10: 99–131. University of Minnesota Press.

Gibbard, Allan, and William Harper. 1978. "Counterfactuals and Two Kinds of Expected Utility." In *Foundations and Applications of Decision Theory*, edited by James L. Leach, Clifford Alan Hooker, and Edward Francis McClennan, 125–62. Reidel.

Goldman, Alvin. 1979. "What Is Justified Belief?" In *Justification and Knowledge*, edited by George Pappas, 1–23. Reidel.

Goldman, Alvin. 1986. *Epistemology and Cognition*. Harvard University Press.

Goldman, Alvin. 2009. "Williamson on Knowledge and Evidence." In *Williamson on Knowledge*, edited by Patrick Greenough and Duncan Pritchard, 74–91. Oxford University Press.

Goodman, Nelson. 1979. *Fact, Fiction and Forecast*. Harvard University Press.

Greco, Daniel. 2015a. "Iteration Principles in Epistemology I: Arguments for." *Philosophy Compass* 10 (11): 754–64.

Greco, Daniel. 2015b. "Iteration Principles in Epistemology II: Arguments against." *Philosophy Compass* 10 (11): 765–71.

Greenspan, Patricia. 1975. "Conditional Ought and Hypothetical Imperatives." *Journal of Philosophy* 72: 259–76.

Hájek, Alan. 2003. "What Conditional Probability Could Not Be." *Synthese* 137: 273–323.

Hájek, Alan. 2008a. "Arguments for—or Against—Probabilism?" *British Journal for the Philosophy of Science* 59 (4): 793–819.

Hájek, Alan. 2008b. "Dutch Book Arguments." In *The Oxford Handbook of Rational and Social Choice*, edited by Prasanta Pattanaik, Paul Anand, and Clemens Pupe, 173–95. Oxford University Press.

Hájek, Alan. 2011. "Conditional Probability." In *Philosophy of Statistics*, Volume 7 (Handbook of the Philosophy of Science), edited by Prasanta S. Bandyopadhyay and Malcolm Forster, 99–135. Elsevier.

Hájek, Alan. ms. "Staying Regular?"

Hansson, Bengt. 1969. "An Analysis of Some Deontic Logics." *Noûs* 3: 373–98.

Harman, Gilbert. 1986. *Change in View: Principles of Reasoning*. MIT Press.

Hawthorne, John. 2002. "Deeply Contingent a Priori Knowledge." *Philosophy and Phenomenological Research* 65 (2): 247–69.

Hawthorne, John. 2004a. *Knowledge and Lotteries*. Oxford University Press.

Hawthorne, John. 2004b. "The Case for Closure." In *Contemporary Debates in Epistemology*, edited by Matthias Steup, 40–55. Blackwell.

Heim, Irene. 1992. "Presupposition Projection and the Semantics of Attitude Verbs." *Journal of Semantics* 9: 183–221.

Hinton, J. M. 1967. "Visual Experiences." *Mind* 76: 217–27.

Huemer, Michael. 2001a. *Skepticism and the Veil of Perception*. Rowman; Littlefield.

Huemer, Michael. 2001b. "The Problem of Defeasible Justification." *Erkenntnis* 54: 375–97.
Ichikawa, Jonathan. 2014. "Justification Is Potential Knowledge." *Canadian Journal of Philosophy* 44 (2): 184–206.
Ismael, Jennan. 2011. "A Modest Proposal about Chance." *Journal of Philosophy* 108 (8): 416–42.
Jeffrey, Richard. 1965. *The Logic of Decision*. 2nd edition. McGraw Hill.
Joyce, James. 1998. "A Non Pragmatic Vindication of Probabilism." *Philosophy of Science* 65 (4): 575–603.
Kaplan, Mark. 1996. *Decision Theory as Philosophy*. Cambridge University Press.
Karttunen, Lauri. 1974. "Presuppositions and Linguistic Context." *Theoretical Linguistics* 1: 181–94.
Kelly, Thomas. 2010. "Peer Disagreement and Higher Order Evidence." In *Disagreement*, edited by Richard Feldman and Ted Warfield, 183–217. Oxford University Press.
Kiesewetter, Benjamin. 2017. *The Normativity of Rationality*. Oxford University Press.
Klein, Peter. 1995. "Skepticism and Closure: Why the Evil Genius Argument Fails." *Philosophical Topics* 23 (1): 213–36.
Kolodny, Niko. 2005. "Why Be Rational?" *Mind* 114 (455): 509–63.
Kolodny, Niko, and John MacFarlane. 2010. "Ifs and Oughts." *Journal of Philosophy* 107 (3): 115–43.
Kotzen, Matthew. forthcoming. "A Formal Account of Epistemic Defeat." In *Themes from Klein*, edited by Branden Fitelson, Rodrigo Borges, and Cherie Branden. Springer.
Kratzer, Angelika. 1991a. "Conditionals." In *Papers from the Parasession on Pragmatics and Grammatical Theory*, edited by Peter Farley, Anne Farley, and Karl McCollough, 1–15. Chicago Linguistics Society.
Kratzer, Angelika. 1991b. "Modality." In *Handbuch Semantik/Handbook Semantics*, edited by Stechow and Wunderlich, 639–50. De Gruyter.
Kyburg, Henry. 1961. *Probability and the Logic of Rational Belief*. Wesleyan University Press.
Lange, Marc. 2000. "Is Jeffrey Conditionalization Defective by Virtue of Being Non-Commutative? Remarks on the Sameness of Sensory Experiences." *Synthese* 123: 393–403.
Lasonen-Aarnio, Maria. 2008. "Single Premise Deduction and Risk." *Philosophical Studies* 141 (2): 157–73.
Lasonen-Aarnio, Maria. 2010. "Unreasonable Knowledge." *Philosophical Perspectives* 24 (1): 1–21.
Lasonen-Aarnio, Maria. forthcoming. "Virtuous Failure and Victims of Deceit." In *The New Evil Demon*, edited by Fabian Dorsch and Julien Dutant. Oxford University Press.
Lewis, David. 1973. *Counterfactuals*. Blackwell.
Lewis, David. 1974. "Semantic Analyses for Dyadic Deontic Logics." In *Logical Theory and Semantic Analysis*, edited by Soeren Stenlund, 1–14. Dordrecht.
Lewis, David. 1975. "Adverbs of Quantification." In *Formal Semantics of Natural Language*, edited by Edward Keenan, 178–88. Cambridge University Press.
Lewis, David. 1979. "Counterfactual Dependence and Time's Arrow." *Noûs* 13: 455–76.
Lewis, David. 1980. "A Subjectivist's Guide to Objective Chance." In *Studies in Inductive Logic and Probability*, edited by Richard Jeffrey, 263–93. University of California Press.
Lewis, David. 1983. "New Work for a Theory of Universals." *Australasian Journal of Philosophy* 61 (4): 343–77.
Lewis, David. 1986. *On the Purality of Worlds*. Blackwell.
Loewer, Barry, and Marvin Belzer. 1983. "Dyadic Deontic Detachment." *Synthese* 54: 295–318.

Lord, Errol. 2018. *The Importance of Being Rational.* Oxford University Press.
McDaniel, Kris, and Ben Bradley. 2008. "Desires." *Mind* 177: 267–302.
McDowell, John. 1982. "Criteria, Defeasibility and Knowledge." *Proceedings of the British Academy* 68: 455–79.
MacFarlane, John. 2010. "Relativism and Knowledge Attributions." In *The Routledge Companion to Epistemology*, edited by Sven Bernecker and Duncan Pritchard, 536–44. Routledge.
Martin, C. B. 1994. "Dispositions and Conditionals." *The Philosophical Quarterly* 44: 1–8.
McKitrick, Jennifer. 2003. "A Case for Extrinsic Dispositions." *Australasian Journal of Philosophy* 81 (2): 155–74.
Meacham, Christopher, and Jonathan Weisberg. 2011. "Representation Theorems and the Foundations of Decision Theory." *Australasian Journal of Philosophy* 89: 641–63.
Moss, Sarah. 2018. *Probabilistic Knowledge.* Oxford University Press.
Mott, Peter. 1973. "On Chisholm's Paradox." *Journal of Philosophical Logic* 2: 197–210.
Nozick, Robert. 1981. *Philosophical Explanations.* Harvard University Press.
Parfit, Derek. 2011. *On What Matters.* Oxford University Press.
Pettigrew, Richard. 2016. *Accuracy and the Laws of Credence.* Oxford University Press.
Pettigrew, Richard. forthcoming. "What Is Justified Credence?", *Episteme.*
Pollock, John. 1970. "The Structure of Epistemic Justification." *American Philosophical Quarterly* 4: 62–78.
Pollock, John. 1974. *Knowledge and Justification.* Princeton University Press.
Pollock, John. 1986. *Contemporary Theories of Knowledge.* 1st edition. Rowman; Littlefield.
Pollock, John, and Joseph Cruz. 1999. *Contemporary Theories of Knowledge.* 2nd edition. Rowman; Littlefield.
Popper, Karl. 1959. *The Logic of Scientific Discovery.* Harper; Row.
Prakket, Henry, and Marek Sergot. 1997. "Dyadic Deontic Logic and Contrary-to-Duty Obligations." In *Defeasible Deontic Logic*, edited by Donald Nute, 223–62. Dordrecht.
Prior, Arthur. 1954. "The Paradoxes of Derived Obligation." *Mind* 63: 64–5.
Pryor, James. 2000. "The Skeptic and the Dogmatist." *Noûs* 34: 517–49.
Pryor, James. 2004. "What's Wrong with Moore's Argument?" *Philosophical Issues* 14: 349–78.
Pryor, James. 2013a. "Problems for Credulism." In *Seemings and Justification: New Essays on Dogmatism and Phenomenal Conservatism*, edited by Chris Tucker, 89–134. Oxford University Press.
Pryor, James. 2013b. "Reply to Comesaña." In *Contemporary Debates in Epistemology*, edited by John Turri Matthias Steup and Ernest Sosa, 2nd edition, 235–8. Blackwell.
Pryor, James. 2013c. "There Is Immediate Justification." In *Contemporary Debates in Epistemology*, edited by John Turri Matthias Steup and Ernest Sosa, 2nd edition, 202–21. Blackwell.
Pryor, James. 2018. "The Merits of Incoherence." *Analytic Philosophy* 59 (1): 112–41.
Putnam, Hillary. 1980. "Models and Reality." *Journal of Symbolic Logic* 45: 464–82.
Quine, Willard van Orman. 1950. *Methods of Logic.* Harvard University Press.
Rayo, Agustin. 2019. *On the Brink of Paradox.* MIT Press.
Ramsey, Frank. 1926. "Truth and Probability." Reprinted in Ramsey, *The Foundations of Mathematics and Other Logical Essays*, 1931, edited by R. B. Braithwaite, 156–98. Kegan, Paul, Trench, Trubner & Co.; Harcourt, Brace and Company.
Ramsey, Frank. 1931. "Knowledge." In *The Foundations of Mathematics and Other Logical Essays*, edited by R. B. Braithwaite, 258–9. Kegan, Paul, Trench, Trubner & Co.; Harcourt, Brace and Company.
Reagan, Donald. 1980. *Utilitarianism and Co-Operation.* Oxford University Press.

Renyi, A. 1970. *Foundations of Probability*. Holden Day Inc.
Rosen, Gideon. 2001. "Nominalism, Naturalism, Epistemic Relativism." *Philosophical Perspectives* 15: 60–91.
Ross, Jacob. 2012. "Rationality, Normativity, and Committment." In *Oxford Studies in Metaethics*, edited by Russ Shaffer-Landau, 7: 138–81. Oxford University Press.
Salerno, Joe. 2009. "Knowability Noir: 1945-1963." In *New Essays on the Knowability Paradox*, edited by Joe Salerno, 29–48. Oxford University Press.
Savage, Leonard. 1954. *The Foundations of Statistics*. John Wiley; Sons.
Scanlon, T. M. 2000. *What We Owe to Each Other*. Harvard University Press.
Schechter, Joshua. 2013. "Rational Self-Doubt and the Failure of Closure." *Philosophical Studies* 163 (2): 428–52.
Schoenfield, Miriam. 2014. "Permission to Believe: Why Permisivism Is True and What it Tells Us about Irrelevant Influences on Belief." *Noûs* 48 (2): 193–218.
Schroeder, Mark. 2007. *Slaves of the Passions*. Oxford University Press.
Schroeder, Mark. 2008. "Having Reasons." *Philosophical Studies* 139 (1): 57–71.
Schroeder, Mark. forthcoming. *Reasons First*. Oxford University Press.
Schwarz, Wolfgang. 2018. "Imaginary Foundations." *Ergo* 5 (29): 764–89.
Sharon, Assaf, and Levi Spectre. 2017. "Evidence and the Openness of Knowledge." *Philosophical Studies* 174 (4): 1001–37.
Shogenji, Tomoji. 2012. "The Degree of Epistemic Justification and the Conjunction Fallacy." *Synthese* 148 (1): 29–48.
Shope, Robert. 1978. "The Conditional Fallacy in Contemporary Epistemology." *The Journal of Philosophy* 75 (8): 397–413.
Smithies, Declan. 2011. "Moore's Paradox and the Accessibility of Justification." *Philosophy and Phenomenological Research* 85 (2): 273–300.
Snowdon, P. F. 1980–1. "Perception, Vision and Causation." *Proceedings of the Aristotelian Society* 81: 175–92.
Sober, Elliott. 2010. "Evolutionary Theory and the Reality of Macro Probabilities." In *The Place of Probability in Science*, edited by Ellery Eells and J. Fetzer, 133–61. Springer.
Sosa, Ernest. 1991. *Knowledge in Perspective*. Cambridge University Press.
Sosa, Ernest. 2016. "Process Reliabilism and Virtue Epistemology." In *Goldman and His Critics*, edited by Brian McLaughlin and Hilary Kornblith, 125–48. Wiley-Blackwell.
Stalnaker, Robert. 1984. *Inquiry*. Bradford Books, MIT Press.
Stalnaker, Robert. 1991. "The Problem of Logical Omniscience, I." *Synthese* 89 (3): 425–40.
Stalnaker, Robert. 1999. "The Problem of Logical Omniscience, Ii." In *Context and Content*, 255–73. Oxford University Press.
Stephánsonn, Orri. 2017. "What Is 'Real' in Probabilism?" *Australasian Journal of Philosophy* 95: 573–87.
Stich, Stephen. 1978. "Autonomous Psychology and the Belief-Desire Thesis." *The Monist* 61 (4): 573–91.
Street, Sharon. 2006. "A Darwinian Dilemma for Realists Theories of Value." *Philosophical Studies* 127: 109–66.
Sturgeon, Scott. 2008. "Reason and the Grain of Belief." *Noûs* 42 (1): 139–65.
Sturgeon, Scott. 2014. "Pollock on Defeasible Reasoning." *Philosophical Studies* 169 (1): 105–18.
Sutton, Jonathan. 2005. "Stick to What You Know." *Nous* 39 (3): 359–96.
Tal, Eyal, and Juan Comesaña. 2015. "Is Evidence of Evidence Evidence?" *Noûs* 51 (1): 95–112.
Tang, Weng Hong. 2016a. "Reliabilism and the Suspension of Belief." *Australasian Journal of Philosophy* 94 (2): 362–77.

Tang, Weng Hong. 2016b. "Reliability Theories of Justified Credence." *Mind* 125 (497): 63–94.

Titelbaum, Michael. 2010. "*Not Enough There There* Evidence, Reasons, and Language Independence." *Philosophical Perspectives* 24: 477–528.

Titelbaum, Michael. 2015. "Rationality's Fixed Point." *Oxford Studies in Epistemology* 5: 253–94.

Titelbaum, Michael. forthcoming. *Fundamentals of Bayesian Epistemology*. Oxford University Press.

Turri, John. 2009. "The Ontology of Epistemic Reasons." *Noûs* 43 (3): 490–512.

van Fraassen, Bas. 1984. "Belief and the Will." *Journal of Philosophy* 81: 235–56.

Vogel, Jonathan. 2000. "Reliabilism Leveled." *The Journal of Philosophy* 97 (11): 602–23.

Vogel, Jonathan. 2014. "E & ~H." In *Scepticism and Perceptual Justification*, edited by Dylan Dodd and Elia Zardini, 88–107. Oxford University Press.

von Fintel, Kai. 2012. "Subjunctive Conditionals." In *The Routledge Companion to Philosophy of Language*, edited by Gillian Russell and Delia Graf Fara, 466–77. Routledge.

von Neumann, John, and Oskar Morgenstern. 1944. *Theory of Games and Economic Behavior*. Princeton University Press.

Weatherson, Brian. 2012. "Knowledge, Bets, and Interests." In *Knowledge Ascriptions*, edited by Jessica Brown and Mikkel Gerken, 75–103. Oxford University Press.

Wedgwood, Ralph. 2013. "A Priori Bootstrapping." In *The a Priori in Philoophy*, edited by Albert Casullo and Joshua Thurow, 226–46. Oxford University Press.

Weisberg, Jonathan. 2009. "Commutativity or Holism? A Dilemma for Conditionalizers." *British Journal for the Philosophy of Science* 60 (4): 793–812.

Weisberg, Jonathan. 2010. "Bootstrapping in General." *Philosophy and Phenomenological Research* 81 (3): 525–48.

Weisberg, Jonathan. 2015. "Updating, Undermining, and Independence." *British Journal for the Philosophy of Science* 66 (1): 121–59.

White, Roger. 2005. "Epistemic Permisiveness." *Philosophical Perspectives* 19 (1): 445–59.

White, Roger. 2006. "Problems for Dogmatism." *Philosophical Studies* 131 (3): 525–57.

White, Roger. 2009. "Evidential Symmetry and Mushy Credence." In *Oxford Studies in Epistemology*, edited by Tamar Szabo Gendler and John Hawthorne, 161–86. Oxford University Press.

Williamson, Timothy. 2000. *Knowledge and Its Limits*. Oxford University Press.

Williamson, Timothy. 2005. "Contextualism, Subject-Sensitive Invariantism, and Knowledge of Knowledge." *The Philosophical Quarterly* 55 (219): 213–35.

Williamson, Timothy. 2009. "Replies to Critics." In *Williamson on Knowledge*, edited by Patrick Greenough and Duncan Pritchard, 279–384. Oxford University Press.

Williamson, Timothy. 2013a. "Gettier Cases in Epistemic Logic." *Inquiry* 56 (1): 1–14.

Williamson, Timothy. 2013b. "Response to Cohen, Comesaña, Goodman, Nagel, and Weatherson on Gettier Cases in Epistemic Logic." *Inquiry* 56 (1): 77–96.

Williamson, Timothy. 2017b. "Ambiguous Rationality." *Episteme* 14 (3): 263–74.

Williamson, Timothy. forthcoming. "Justification, Excuses and Sceptical Scenarios." In *The New Evil Demon*, edited by Julien Dutant and Fabian Dorsch. Oxford University Press.

Wright, Crispin. 2004. "Warrant for Nothing (and Foundations for Free)?" *Aristotelian Society Supplementary Volume* 78 (1): 167–212.

Wright, George Henrik von. 1956. "A Note on Deontic Logic and Derived Obligation." *Mind* 65: 507–9.

Yablo, Stephen. 1999. "Intrinsicness." *Philosophical Topics* 26: 590–627.

Zynda, Lyle. 2000. "Representation Theorems and Realism about Degrees of Belief." *Philosophy of Science* 67: 45–69.

Index

Alston, William 176, 201, 203
Accuracy argument 2, 19, 21–27, 37, 41,
Ampliativity, principle 178, 182, 185–187, 189
Armendt, Brad 21
Arntzenius, Frank 66

Bad Lucas case 87–91, 94–97, 115, 121
Ball, Brian 113, 206
Bayesianism 3, 41, chapter 3, 70, 73, 77, 82, 84, 93, 133, 140, 142, 146, 192–193, 212
Bird, Alexander 107, 112
Bloome-Tillman, Michael 113, 206
BonJour, Lawrence 198, 205, 207
Bronfman, Aaron 172
Broome, John 5, 148–152, 162, 163
Byrne, Alex 123

Carnap, Rudolph 12, 43, 53–59, 62, 64, 203, 210
Chisholm, Roderick 4, 153–161, 165, 167, 172, 175–176
Christensen, David 21, 29–30, 61
Conee, Earl 118, 193–198
Closure 149–152, 155, 158, 161–163, 178–183, 186–187, 189
Cohen, Stewart 16, 22, 71, 83, 85, 91, 92, 99, 100, 113, 134, 173, 175–177, 186, 206, 207
Conditionalization 15–16, 31–33, 46–49, 65–67
–, Jeffrey 141–142, 146
–, ur-prior 65–67, 145–146
Conditionals, restrictor view of 167–168
Contrary-to-duty obligations 150, 153–156, 161, 163, 166–167, 169, 172

Dancy, Johnathan 1, 69, 118, 119, 212
Defeaters 120, 134–146, 176
Deontic logic 4, 150–151, 153, 155, 157, 160–161, 163, 165, 169, 172
–, dyadic 4, 157, 160–161, 163, 165–170, 172
Dilemmas 72–77, 95–97, 128–129
Dowell, Janice 152, 171–172
Dutant, Julien 3, 88
Dutch Book argument 2, 19–22, 26–27, 41

E = K *see* factualism
Entailment, principle 181–182, 184–187, 189–191

Evidence 30, 32–33, 44, 47–49, 51–75, 82–86, 91–93, 99, 102, 110–113, 115, 116–127, 130–135, 139–146, 149, 150, 171–183, 178, 185–198, 201–203, 205–212
Evidentialism 4–5, 193–197, 209–210
Evidentialist Reliabilism 4, 193, 203, 208–212
Excuses 3, 5, 93, chapter 5, 129
Experientialism 2–4, 116–124, 126–127, 129–135, 139–140, 145–146, 189–191, 211–212

Factualism 1–5, 68–71, 82–86, 92, 117–124, 128–130, 173–174, 187, 191–192, 196, 211–212
Fantl, Jeremy 16, 85, 212
Feldman, Richard 59, 118, 193–198
Fitelson, Branden 12, 53–54, 58
Fumerton, Richard 3–4, 71, 73–77, 81, 92–94, 97, 128, 146, 148, 150, 152, 173–176, 209
Fundamental Theorem of Bayesian Epistemology 49, 50

Goldman, Alan 75, 85, 119, 132–133, 176, 197–198, 204, 210
Good Lucas case 87, 89–91, 96, 121
Goodman, Nelson 43, 57, 59, 62–64
Grue 43, 48, 49, 57–59, 62–65

Hájek, Alan 13, 14, 20, 24, 79–80, 85–86
Hawthorne, John 16, 65, 179
Huemer, Michael 175–177, 182

Ichikawa, Johnathan 107

Jeffrey, Richard 34, 38, 141–142, 145–146

Kolodny, Niko 152, 156, 166–168, 171–172
Kratzer, Angelika 153, 168–169, 171–172

Lasonen-Aarnio, María 95, 100–107, 109–113, 179–180, 182–183
Lewis, David 9, 14, 32, 43, 50–51, 63–64, 153, 155, 157, 160, 164, 168
Lucas 1–2, 87–92, 94–97, 115, 121, 131–132, 211

MacFarlane, John 16, 152, 156, 167–168, 171–172
McGrath, Matthew 16, 85, 119, 131, 144, 212
Mere Lemmas, principle 180–183, 186–187, 189–191, 195, 208, 211
Modus Ponens 169–171

Parfit, Derek 73, 87, 116–117, 128, 148, 167
Pettigrew, Richard 4, 22, 193, 204–207, 210
Pollock, John 118, 120, 127, 134–141, 144, 146, 173, 175–176
Principle of Indifference 52–53, 55–56
Probabilism 16, 19–49
Pryor, James 118, 121, 135, 173, 175–177, 179–184
Psychologism 1–4, 118–127, 130, 135, 143–146, 211–212

Ramsey, Frank 20, 197
Reliabilism 4, 22, 100, 113, 173–176, 192–193, 197–212

Schroeder, Mark 119, 134, 136, 212
Sosa, Ernest 175, 208
Stalnaker, Robert 9, 30, 157, 164, 168
Sturgeon, Scott 61, 144

Titelbaum, Michael 2, 43–44, 62–64, 91, 206
Tomás 1–2, 87–92, 94–97, 115, 121, 131–132, 211–212

Ur-prior 15, 44–53, 62–68, 113, 207–208, (*see also* sky)

Vogel, Johnathan 173–176, 184

Weisberg, Johnathan 41, 134, 142–143, 181
White, Roger 59, 61, 135
Williamson, Timothy 44, 66–71, 82–100, 107, 113, 117–118, 122–123, 132–133, 179, 187, 211–212